GAS-SOLID TRANSPORT

McGraw-Hill Chemical Engineering Series

BUILDING THE LITERATURE OF A PROFESSION

Fifteen prominent chemical engineers first met in New York more than 50 years ago to plan a continuing literature for their rapidly growing profession. From industry came such pioneer practitioners as Leo H. Baekeland, Arthur D. Little, Charles L. Reese, John V. N. Dorr, M. C. Whitaker, and R. S. McBride. From the universities came such eminent educators as William H. Walker, Alfred H. White, D. D. Jackson, J. H. James, Warren K. Lewis, and Harry A. Curtis. H. C. Parmelee, then editor of *Chemical and Metallurgical Engineering,* served as chairman and was joined subsequently by S. D. Kirkpatrick as consulting editor.

After several meetings, this committee submitted its report to the McGraw-Hill Book Company in September 1925. In the report were detailed specifications for a correlated series of more than a dozen texts and reference books which have since become the McGraw-Hill Series in Chemical Engineering and which became the cornerstone of the chemical engineering curriculum.

From this beginning there has evolved a series of texts surpassing by far the scope and longevity envisioned by the founding Editorial Board. The McGraw-Hill Series in Chemical Engineering stands as a unique historical record of the development of chemical engineering education and practice. In the series one finds the milestones of the subject's evolution: industrial chemistry, stoichiometry, unit operations and processes, thermodynamics, kinetics, and transfer operations.

Chemical engineering is a dynamic profession, and its literature continues to evolve. McGraw-Hill and its consulting editors remain committed to a publishing policy that will serve, and indeed lead, the needs of the chemical engineering profession during the years to come.

THE SERIES

Bailey and Ollis: *Biochemical Engineering Fundamentals*
Bennett and Myers: *Momentum, Heat, and Mass Transfer*
Beveridge and Schechter: *Optimization: Theory and Practice*
Carberry: *Chemical and Catalytic Reaction Engineering*
Churchill: *The Interpretation and Use of Rate Data – The Rate Concept*
Clarke and Davidson: *Manual for Process Engineering Calculations*
Coughanowr and Koppel: *Process Systems Analysis and Control*
Danckwerts: *Gas Liquid Reactions*
Finlayson: *Nonlinear Analysis in Chemical Engineering*
Gates, Katzer, and Schuit: *Chemistry of Catalytic Processes*
Harriott: *Process Control*
Holland: *Fundamentals of Multicomponent Distillation*
Johnson: *Automatic Process Control*
Johnstone and Thring: *Pilot Plants, Models, and Scale-up Methods in Chemical Engineering*
Katz, Cornell, Kobayashi, Poettmann, Vary, Elenbaas, and Weinaug: *Handbook of Natural Gas Engineering*
King: *Separation Processes*
Klinzing: *Gas-Solid Transport*
Knudsen and Katz: *Fluid Dynamics and Heat Transfer*
Lapidus: *Digital Computation for Chemical Engineers*
Luyben: *Process Modeling, Simulation, and Control for Chemical Engineers*
McCabe and Smith, J. C.: *Unit Operations of Chemical Engineering*
Mickley, Sherwood, and Reed: *Applied Mathematics in Chemical Engineering*
Nelson: *Petroleum Refinery Engineering*
Perry and Chilton (Editors): *Chemical Engineers' Handbook*
Peters: *Elementary Chemical Engineering*
Peters and Timmerhaus: *Plant Design and Economics for Chemical Engineers*
Ray: *Advanced Process Control*
Reed and Gubbins: *Applied Statistical Mechanics*
Reid, Prausnitz, and Sherwood: *The Properties of Gases and Liquids*
Resnick: *Process Analysis and Design for Chemical Engineers*
Satterfield: *Heterogeneous Catalysis in Practice*
Sherwood, Pigford, and Wilke: *Mass Transfer*
Slattery: *Momentum, Energy, and Mass Transfer in Continua*
Smith, B. D.: *Design of Equilibrium Stage Processes*
Smith, J. M.: *Chemical Engineering Kinetics*
Smith, J. M., and Van Ness: *Introduction to Chemical Engineering Thermodynamics*
Thompson and Ceckler: *Introduction to Chemical Engineering*
Treybal: *Mass Transfer Operations*
Van Winkle: *Distillation*
Volk: *Applied Statistics for Engineers*
Walas: *Reaction Kinetics for Chemical Engineers*
Wei, Russell, and Swartzlander: *The Structure of the Chemical Processing Industries*
Whitwell and Toner: *Conservation of Mass and Energy*

GAS-SOLID TRANSPORT

George E. Klinzing

Chemical/Petroleum Engineering Department
University of Pittsburgh

McGraw-Hill Book Company

New York   St. Louis   San Francisco   Auckland   Bogotá   Hamburg
Johannesburg   London   Madrid   Mexico   Montreal   New Delhi   Panama
Paris   São Paulo   Singapore   Sydney   Tokyo   Toronto

To THE MEMORY OF MY FATHER
Engelbert Klinzing

6385 - 7078

CHEMISTRY

This book was set in Press Roman by Meredythe.
The editors were Diane D. Heiberg and Ken Burke;
the production supervisor was Diane Renda.
R. R. Donnelley & Sons Company was printer and binder.

GAS-SOLID TRANSPORT

1 2 3 4 5 6 7 8 9 0 D O D O 8 9 8 7 6 5 4 3 2 1

Library of Congress Cataloging in Publication Data

Klinzing, George E
    Gas-solid transport.

    (McGraw-Hill chemical engineering series)
    Includes bibliographical references and index.
    1. Transport theory. 2. Solids. I. Title.
TP156.T7K55   660.2'8423   80-23068
ISBN 0-07-035047-7

CONTENTS

PREFACE

The transport of solids is a vital operation in many manufacturing processes. It is difficult to imagine an industry that does not concern itself with solids usage in some way.

Bulk handling of solids is usually accomplished by use of bins and conveyors. But solids may also be transported by suspension in a liquid or a gas. The latter method of transport is the subject of this book. Often, the suspension of solids in a liquid is not desirable from a process standpoint. The solids may dissolve or change character in the liquid, or the presence of the liquid may not be desirable in some high-temperature process. In these cases, one must resort to gas-solid transport or pneumatic transport of the solids for movement.

The implementation of gas-solid transport involves several technological problems. Many challenging areas of research remain that may make direct impacts on industry. The area of solids handling and, in particular, gas-solid transport is very important in the synthetic energy production field and in the ceramics field. Almost all synfuel processes start with a solid feed such as coal, oil shale, wood, waste, etc. To deliver these materials into a liquid or gaseous fuel, these solids and their by-products must be transported to, through, around, and from the process. The use of high pressures and highly dense flow systems is finding increased importance. Our knowledge in these areas is not strong.

As the sizes of the particles of solids are reduced, the surface area of the material increases and its reactivity is significantly augmented. Fine sizing is often the correct procedure for a process, but with the production of fine particles the handling and transport difficulties increase dramatically. When dealing with fine particulate matter, the true particle size for a process is often unknown. Agglomeration and adhesion forces come into play for fine powder and often dominate the situation. Questions such as what is the appropriate particle size for processing, how much material adheres to the surface, does the particulate matter agglomerate, how does one mix two solid powders, what are the energy losses on transporting solids in a gas stream, how can one accurately measure gas-solid flows, and can one expect electrostatics to play a significant role in gas-solid flow systems are often posed in designing gas-solid systems. This book will attempt to answer these questions, giving guidelines and specific values for the final design.

The book is aimed at a specialized course in the gas-solid handling area or the advanced undergraduate or graduate level. A number of suggested problems are included at the end of each chapter and example problems are given in the text. Current developments and research in the area are noted.

Several students and colleagues have contributed much to the preparation of this book. I am particularly indebted to former and present students, especially Moonis Ally, Dan Bender, Len Peters, and Mark Weaver. My associates in government and industry have been most helpful in providing me with proper practicality. For this assistance I would like to acknowledge Dan Bienstock, Jim Joubert, Mahendra Mathur, and Gene Smeltzer. In preparing the final manuscript Rita Sullivan has typed, edited, and suggested uniformity in the text. I wish to express special thanks to her and her ability. For checks on numerical examples, Richard Shuck has been of great assistance. My wife Sandy has proofed the manuscript efficiently, translating non-English to English, and tolerating my bad temper during some of the tedium involved in preparation of such a book.

George E. Klinzing

LIST OF SYMBOLS

A = area

A^* = parameter in Cunningham correction factor.

A_{2_1} = area of contact between particles 1 and 2

A' = Hamaker constant

A_c = contact area

A_{c_e} = elastic contact area

A_{c_p} = plastic contact area

A_p = projected area

a_c = radius of contact

a_q = radius of impacting particle

B = constant of proportionality

C = capacitance

C_c = circumference

C_D = drag coefficient

C_{DM} = drag coefficient of mixture

C_{D_s} = drag coefficient of single particle

$C_t = \dfrac{\text{solid volumetric flow rate}}{\text{total volumetric flow rate}}$

D = diameter of pipe; diameter of particle

$\bar{D}$ = arithmetic mean diameter

$\underline{D}$ = major screw diameter

D_a = diameter of a circle having the same projected area as the particle in a stable position

D_c = cloud diameter

D_g = log geometric mean diameter

D_l = length mean diameter

D_o = orifice diameter

D_p = particle diameter

D_s = surface mean diameter

D_v = volume mean diameter

D_{vs} = volume-surface mean diameter

D_w = weight mean diameter

$\mathscr{D}$ = electric flux density

$\mathscr{D}_s$ = density per unit energy of surface levels

$$\underline{d} = \text{minor screw diameter}$$
$$d_1, d_2 = \text{characteristic charge-transfer lengths}$$
$$E = \text{electric field}$$
$$E_0 = \text{induction constant}$$
$$E_c = \text{bottom edge of conduction band energy}$$
$$E_{F_s} = \text{energy of Fermi level}$$
$$E_i = \text{elastic modulus}$$
$$E_v = \text{top edge of valence band energy}$$
$$e = \text{electronic charge}$$
$$F = \text{force}$$
$$F' = \text{flow rate}$$
$$F_{ad} = \text{adhesion force}$$
$$F_{cap} = \text{capillary force}$$
$$F_e = \text{external force}$$
$$F_{el} = \text{electrostatic force}$$
$$F_{IFT} = \text{interfacial tension force}$$
$$F_L = \text{Magnus force}$$
$$F_{LP} = \text{liquid pressure force}$$
$$Fr = \text{Froude number}$$
$$Fr_0 = \text{Froude number based on terminal velocity}$$
$$F_r = \text{radial drag force}$$
$$F_t = \text{transverse drag force}$$
$$F_{VDW} = \text{van der Waals force}$$
$$\mathscr{F} = \text{free energy}$$
$$f = \text{single-phase friction factor}$$
$$f_g = \text{fluid friction factor}$$
$$f_m = \text{friction factor of mixture}$$
$$f_{n,m} = \text{particle size distribution}$$
$$f_p = \text{particle friction factor}, f_p = 4f_s$$
$$f_s = \text{solid friction factor}$$
$$G = \text{mean particle weight}$$
$$g = \text{gravitational constant}$$
$$g_r, g_\theta, g_z = \text{gravitational force in } r, \theta, z \text{ directions}$$
$$H = \text{magnetic field}$$
$$H' = \text{height of particle}$$
$$h = \text{transfer coefficient}$$
$$\hbar = \text{Planck's constant}$$
$$\mathbf{i} = \text{unit vector}$$
$$I_R = \text{turbulence intensity}$$
$$i = \text{current}$$
$$k = \text{permeability}$$
$$\mathbf{k} = \text{unit vector}$$
$$k_0 = \text{spring constant}$$

$L, L' = $ length

$\mathbf{L} = $ unit radial vector

$L_{accel} = $ acceleration length

$L_e = $ equivalent length

$l = $ mean free path

$l^* = $ leakage factor

$\bar{l} = $ distance between center of spheres

$l_c = $ length of probe

$M = $ mass

$\dot{M} = $ mass flow rate

$\mathbf{M} = $ unit transverse vector

$M_T = $ total mass of sample

$m' = $ flakeness

$m_p = $ mass of particle

$m_q = $ mass of component q

$m_s = $ mass of solids

$N = $ number density

$\mathbf{N} = $ normal force

$N_{ev} = $ electroviscous number

$N_{Im} = $ impact number

$N_q = $ number density of component q

$N_R = $ rps

$N_T = $ total number of particles

$n = $ number of collisions

$n' = $ elongation

$n^* = $ rebound coefficient

$n_R = $ refractive index

$P = $ pressure

$P' = $ pitch

$\Delta P = $ pressure drop

$P_i = $ mass fraction

$\tilde{p} = $ dipole moment

$p_\gamma = $ capillary pressure

$q = $ charge

$q_0 = $ initial charge on particle

$q_m = $ asymptotic separation charge

$q_s = $ saturation charge

$R = $ radius of particle

$R_B = $ radius of bend

$R_c = $ radius of contact

$R_{CL} = $ distance from center of centrifuge axis

$R_{CY} = $ radius of cylinder

$Re = $ Reynolds number

Re_c = critical Reynolds number

Re_p = particle Reynolds number

Re_x = length Reynolds number

r = radial position

r_0 = radius of pipe

r_0^* = radius of curvature of particle

r_c = radius of probe

$\mathbf{r}_i$ = unit vector of scattered radiation

$\mathbf{r}_s$ = unit vector of incident radiation

S = surface area

T = temperature

T^* = charge leakage constant

T' = torque

$\bar{T}$ = stress tensor

T_c = Tensile strength

t = time

$\underline{t}$ = thickness of flight

Δt = contact time

U = one-phase fluid velocity

U_f = fluid velocity

U_g = superficial gas velocity

U_L = lateral velocity

$U_{\min}$ = minimum velocity

U_p = particle velocity

$\bar{U}_p$ = average velocity of particle

U_{pt} = particle terminal velocity

U_s = saltation velocity

U_s' = superficial solid velocity

U_s^* = friction velocity at saltation

U_{s0}^* = friction velocity at saltation for single particle

U_t = terminal velocity

$\bar{U}_t$ = average terminal velocity

u = fluid velocity in x direction

u_f = fluid velocity in x direction

u_p = particle velocity in x direction

u_q = velocity component in x direction

u_t = transverse velocity of fluid

V = voltage

V = volume

$\bar{V}$ = mean gas velocity

V_c = contact voltage

V_{choke} = choking velocity

V_l = leakage voltage

v = fluid velocity in radial direction
v_0 = characteristic velocity
v_e = velocity of particle at elastic limit
v_f = fluid velocity in radial direction
v_p = particle velocity in radial or perpendicular-to-flow direction
v_p^* = volume of particle
v_q = velocity component in r direction
v_r = radial velocity
v_s = rolling speed; surface separation velocity
v_t = tangential velocity to flow
$(v_1^2)^{1/2}$ = velocity of particles hitting the wall due to random motion
W' = width of particle
W_b = energy difference between Fermi level and conduction band
W_g = fluid flow rate
W_s = solid flow rate
W_ϕ = work function
w = velocity in angular direction
$\bar{w}$ = angular rotation
w_p = particle velocity in angular direction
w_q = velocity component in angular direction
X = loading
x = weight fraction; position
y = weight fraction; position
y^* = yield pressure
y^+ = dimensionless distance $\dfrac{y}{D} \dfrac{Re}{2} \sqrt{\dfrac{f_g}{2}}$
Z_0 = distance
z = position

Greek

α = angle in interfacial forces analysis; geometric angle
β = interfacial tension
γ = interfacial tension
γ' = Euler's constant
Δ = ratio of change in stresses due to the presence of particles divided by stresses normally present
δ = boundary-layer thickness
ϵ = voidage
ϵ_i = dielectric constant
ϵ_f = eddy diffusivity of fluid
ϵ_p = eddy diffusivity of particle
ϵ_s = voidage at saltation

θ = angular position; hit angle

θ_1 = wetting angle

θ_2 = angle of inclination in a bend

κ = permittivity

λ^* = parameter $(5\mu_f/\rho_f U^*)$

λ = drag force modification term

λ_0 = vacuum wavelength of light

μ = viscosity

μ_0 = magnetic permeability

μ_1, μ_2 = Poisson ratio

μ_f = fluid viscosity

μ_s = coefficient of sliding friction

ν = kinematic viscosity

ν_D = Doppler shift

ξ = imaginary frequency

ρ = one-phase density

ρ^* = density ratio ρ_p/ρ_f

ρ_1, ρ_2 = density of particle 1, particle 2

ρ_B = bulk density of solids

ρ_{ds} = cloud density

ρ_f = density of fluid

ρ_m = mixture density

ρ_p = density of particle

ρ_q = volume charge density

ρ_q^* = density of component q based on total volume of mixture

σ = surface charge density; standard deviation

σ' = boundary-layer thickness in electrical adhesion force analysis

σ_1, σ_2 = electrical conductivity of surface 1, surface 2

σ_e = electrical conductivity

σ_g = log geometric standard deviation

σ_i = conductivity

σ_R = theoretical standard deviation

τ = period

τ' = relaxation time

τ_q = relaxation time

ϕ = angle of bend

Φ_T = friction modification term

$\phi(t)$ = contact distribution

φ = intrinsic potential

χ = electron affinity

Ψ = sphericity factor

Ψ' = parameter in Schuchart's bend analysis

ω = frequency

PARTICLE SIZE ANALYSIS

1-1 INTRODUCTION

A basic parameter in all gas-solid systems is the size of the particles present in the system. At first glance, determination of particle size may seem like an easy task. On taking up a handful of sand, inspection by eye tells one that an average particle size could probably be determined with ease. A simple technique would be to sieve the sand into different fractions and weight average the results according to the sieve openings. This technique is probably sufficient for a number of situations, but there are many more where this procedure is not applicable.

In considering particle size it is important to know the state in which the particles exist under the conditions of interest. Are the particles in a static or dynamic situation? Is the distance between adjacent particles intimate or separated sufficiently to allow a dilute phase transfer situation? Is there a high degree of probability of particle-particle interaction or particle-surface interaction? Often, the analysis technique for determining average particle size cannot be performed under actual process conditions, and one must rely on static analysis procedures.

In considering the sand analysis described above, one finds generally little adhesion or agglomeration of the particles with oven-dried sand. Moist sand, of course, is difficult if not impossible to handle by standard sieving procedures, and wet sieving techniques must be employed. If the particulate matter one is considering has adhering or agglomerating tendencies, appropriate sizing is quite difficult. Even if these tendencies can be reduced in static measurements, the question still arises as to whether this state is representative of the actual dynamic situation.

Many times, the size distributions of the particulate matter are not gaussian. Research has been able to produce monodispersed particles to use as standards for comparison of behavior to polydispersed systems, and a number of studies have investigated the effects of varying degrees of polydispersivity. Since it seems unlikely that many industrial processes will be able to afford the luxury of maintaining mono-dispersed systems, polydispersion is a condition worthy of study.

Another concept of critical interest in dealing with solid particles is that of shape. The weight or volume (or some other characteristic dimension) of particles of irregular shape may sometimes be expressed in terms of equivalent spherical particles. This procedure is useful, because it simplifies calculations, but it tends to lead the novice to believe that particles are more often spherical than not. Of course, this is not the case.

With these ideas of size and shape in mind, one can attempt to define the particles quantitatively.

1-2 DISTRIBUTIONS AND AVERAGING

The distribution of particles over varying size ranges usually can be given on a mass or number basis. Sieve analyses lend themselves to a mass basis, while optical sizing techniques are most often on a number basis. The distribution function for a mass distribution can be written as f_m, and the number distribution function can be written as f_n. The mass of particles in a given size range D to $D + dD$ (where D is the diameter of the particle) can be represented as

$$dM = M_T f_m(D) \, dD \tag{1-1}$$

and the number of particles in the same size range as

$$dN = N_T f_n(D) \, dD \tag{1-2}$$

The mass and number distributions can be shown to be related as follows:

$$f_m(D) = \frac{N_T}{M_T} m(D) f_n(D) \tag{1-3}$$

where N_T is the total number of particles, M_T is the total mass of the sample, and $m(D)$ is the mass of particles of size D. Many different types of distributions can be considered for the analyses of particles. The most common distribution is known as the gaussian or normal distribution; however, when dealing with particles the log normal distribution is often found to give the best representation of the data. In the coal-handling area the Rosen–Rammler distribution has found wide acceptance. Table 1-1 gives the expressions for these various distribution functions. Figure 1-1 shows a log normal distribution.

There are other distributions that are possible for particulate matter. One non-linear distribution of interest is the bimodal distribution, representing a division of the particles into two distinct groupings. This bimodal distribution provides some preliminary information, suggesting that such a distribution increases the flowability of a mixture. Allen (1975) has shown that by plotting the bimodal distribution on log probability paper, a change in slope of the line is seen and bimodality is therefore indicated.

Table 1-1 Distribution functions

Type	$f_n(D)$
Gaussian (normal)	$\dfrac{1}{\sigma\sqrt{2\pi}} \ \exp \dfrac{-(D-\bar{D})^2}{2\sigma^2}$
Log normal	$\dfrac{1}{\log \sigma_g\sqrt{2\pi}} \ \exp\dfrac{-(\log D - \log D_g)^2}{2\log^2\sigma_g}$
Rosen–Rammler	$-abD^{b-1}\exp(-aD^b)$

Note: (σ = standard deviation

$$= \left\{ \Sigma\ [N(D-\bar{D})^2]\ /\ \Sigma\ N \right\}^{1/2}$$

$\bar{D}$ = arithmetic mean diameter = $\Sigma\ DN/\Sigma\ N$

$\log \sigma_g$ = log geometric standard deviation

$$\doteq \sqrt{\Sigma\ [N(\log D - \log D_g)^2]/\Sigma\ N}$$

$\log D_g$ = log geometric mean diameter

$$D_g = \sqrt[N]{D_1, D_2, \ldots, D_N}$$

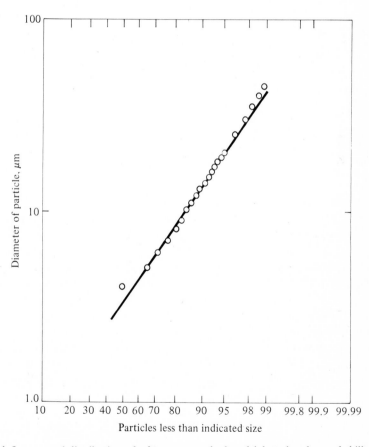

Figure 1-1 Log normal distribution of a Montana rosebud coal (plotted on log probability paper).

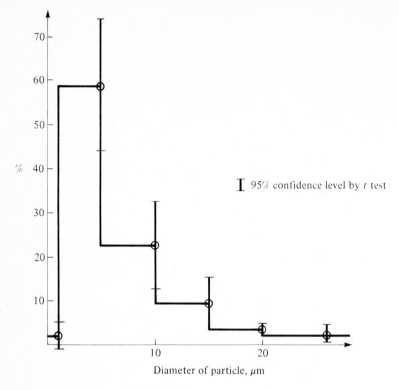

%

Diameter of particle, μm

Figure 1-2 Histogram of pulverized Montana rosebud coal.

In order to construct a bimodal distribution, narrow cuts of large and small particles can be achieved by sieving the usual unimodal distribution samples. These narrow cuts can then be combined to obtain the bimodal distribution. This sieving procedure and distribution tailoring has the potential for industrial application if the narrowness of the sieve cuts is not stringent. Two different settings on industrial pulverizers could also be employed with two pulverizers followed by a blending opera-tion to achieve bimodal character of a sample.

Histograms are a convenient way to represent the distribution of particles. This technique involves the discrete allocation of particles of a certain size according to frequency of occurrence. Figure 1-2 shows the histogram of a coal sample.

Example 1-1 shows the raw data for the particle sizes of a Pittsburgh seam coal sample and the corresponding histogram. Example 1-2 gives the size distribution of a powder with a bimodal distribution.

Example 1-1 A sieve analysis of a Pittsburgh seam coal was performed, yielding the data in Table 1-2. Construct a histogram of the data. (See Fig. 1-3.)

Table 1-2

Diameter μm	Frequency	Diameter μm	Frequency
1	2.62	9	1.5
2	18.7	10	1.8
3	24.3	15	4.9
4	14.9	20	2.1
5	10.2	25	1.1
6	5.2	40	1.6
7	2.9	50	0.27
8	2.0		

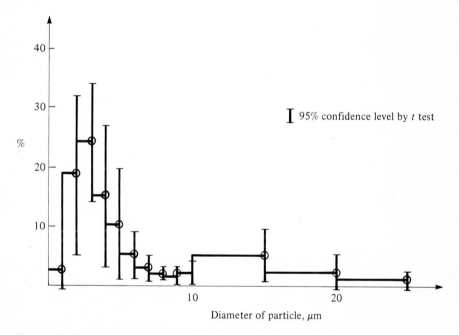

Figure 1-3 Histogram of pulverized Pittsburgh seam coal.

Example 1-2 The data in Table 1-3 are given for a uranium oxide powder as size versus frequency. Plot the data on log probability paper and make conclusions accordingly.

From Fig. 1-4 one can see that a bimodal distribution results for the uranium oxide powder, with peak diameters at 0.6 μm and 2 μm.

A single spherical particle has one diameter than can be measured as being characteristic. If the world were made of these spherical particles, then there would be little

Table 1-3

Cumulative mass % finer than	Δ	Equivalent spherical diameter, μm
100		
	2	30
98		
	2.5	20
95.5		
	1.0	10
94.5		
	2.0	9
92.5		
	2.5	8
90		
	3.5	7
86.5		
	4.0	6
82.5		
	5.0	5
77.5		
	10.0	4
67.5		
	14.5	3
53		
	17	2
36		
	2	1
34		
	3	0.9
31		
	3	0.8
28		
	3	0.7
25		
	2.5	0.6
22.5		
	3.5	0.4
19		
	11	0.3
17		
	2.0	0.2
6		
	6	0.1
0		

ambiguity when it came to determining the correct particle size for solid objects. As we mentioned earlier, the diameter of a particle is usually based on an equivalent sphere of the same volume or mass. This can be expanded also to equivalent surface area. The diameter of an irregular solid can also be related to the dynamic situation of

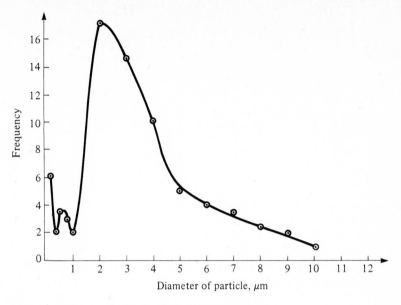

Figure 1-4 Bimodal distribution of uranium oxide powder.

the movement of the object in a fluid. Considering this fluid movement, one can define a drag diameter as the diameter of a sphere having the same resistance to motion as the particle of interest in the same fluid at the same velocity. A free-fall diameter is similarly defined but has the particle at its terminal velocity in the fluid. The Stokes diameter limits this velocity to the laminar flow regime. Because sieve analyses are easy to perform and quite common industrially for characterizing particulate matter, the sieve diameter is of considerable utility. The sieve diameter is the width of the minimum square aperture through which the particle will pass. Microscopic techniques in characterizing particles often use the mean value of the distance between pairs of parallel tangents to the projected outline of particles, the Feret diameter, and the mean chord length of the projected outline of the particle, Martin's diameter.

The dilemma in deciding the appropriate particle size does not stop once a character length has been decided upon. Various forms of averaging are possible. The specific manner of averaging is often dependent upon the process in which the particle is found. When one is dealing with momentum transfer at low velocities, the length mean diameter is probably most appropriate. The weight mean diameter obtained from the sieve analysis is important in the settling area. The volume mean diameter is most relevant for void fraction determination, and optical techniques should be applied for these determinations. In mass and heat transfer problems the surface area is of utmost importance, so the surface mean diameter of particles involved in these processes should be used. For combustion and chemical reactions taking place on particles one should generally employ the volume to surface diameter. This volume-surface mean diameter is also called the Sauter mean diameter and is used extensively in liquid-liquid and liquid-gas processing. Table 1-4 lists the average diameters that may be used for analyses.

Table 1-4 Average diameters

Type	General formula
Length mean diameter	$D_l = \Sigma ND/\Sigma N$
Surface mean diameter	$D_s = (\Sigma ND^2/\Sigma N)^{1/2}$
Volume mean diameter	$D_v = (\Sigma ND^3/\Sigma N)^{1/3}$
Volume surface mean diameter	$D_{vs} = \Sigma ND^3/\Sigma ND^2$
Weight mean diameter	$D_w = \Sigma ND^4/\Sigma ND^3$
Log mean diameters	
length	$D_{lg} = \exp(\log D_g + 0.5 \log^2 \sigma_g)$
surface	$D_{sg} = \exp(\log D_g + 1.0 \log^2 \sigma_g)$
volume	$D_{vg} = \exp(\log D_g + 1.5 \log^2 \sigma_g)$
volume-surface	$D_{vsg} = \exp(\log D_g + 2.5 \log^2 \sigma_g)$
weight	$D_{wg} = \exp(\log D_g + 3.0 \log^2 \sigma_g)$

Note: $\log D_g$ is the log geometric mean diameter and $\log \sigma_g$ is the log geometric standard deviation.

The general expressions given in Table 1-4 for the mean diameters may also be written in terms of the distribution functions as

$$D_l = \frac{\int D f_n \, dD}{\int f_n \, dD} \qquad (1\text{-}4)$$

$$D_s = \left(\frac{\int D^2 f_n \, dD}{\int f_n \, dD}\right)^{1/2} \qquad (1\text{-}5)$$

$$D_v = \left(\frac{\int D^3 f_n \, dD}{\int f_n \, dD}\right)^{1/3} \qquad (1\text{-}6)$$

$$D_w = \frac{\int D^4 f_n \, dD}{\int D^3 f_n \, dD} \qquad (1\text{-}7)$$

$$D_{vs} = \frac{\int D^3 f_n \, dD}{\int D^2 f_n \, dD} \qquad (1\text{-}8)$$

Any distribution function that represents the particle sample can be inserted in the above to determine the average properties. Insertion of the normal distribution function in these expressions yields the forms for the averages given by equations in Table 1-4. Using the log normal distribution, additional averages can be found on this basis.

The various diameter averages are given for two different coal types in Examples 1-3 and 1-4.

Example 1-3 The optical microscope analysis given in Table 1-5 was performed for a particulate sample of ground Pittsburgh seam coal that was to be fed to a combustor. Determine the length, surface, volume, and weight mean diameters. Using the equations in Table 1-4, one finds

$$D_l = 11.85 \ \mu m$$
$$D_s = 20.15 \ \mu m$$

$$D_v = 33.16 \,\mu\text{m}$$
$$D_w = 141.1 \,\mu\text{m}$$

Table 1-5

Diameter, μm	Cumulative number	Diameter, μm	Cumulative number
0	524	14	120
1	510	15	109
2	495	20	72
3	470	25	43
4	424	30	34
5	361	35	24
6	314	40	14
7	269	45	10
8	243	50	9
9	217	60	8
10	189	70	6
11	172	85	4
12	140	100	2
13	136	125	2

Example 1-4 A ground Montana rosebud coal was found to follow the log normal distribution. Since this is so, the mean diameter averaging is suggested to be performed on this basis. The size distribution is given in Table 1-6. Determine the length, surface, volume, and weight averages based on the log normal distribution. Using the equations in Table 1-4 again, one finds

$$D_{lg} = 7.12 \,\mu\text{m}$$
$$D_{sg} = 9.29 \,\mu\text{m}$$
$$D_{vg} = 12.09 \,\mu\text{m}$$
$$D_{wg} = 15.83 \,\mu\text{m}$$

Table 1-6

Diameter, μm	Cumulative number	Diameter, μm	Cumulative number
0	861	10	268
1	831	11	132
2	737	12	111
3	680	13	95
4	469	14	81
5	430	15	68
6	316	20	28
7	267	25	9
8	239	30	4
9	201	40	1

1-3 SHAPE

Shape is a characteristic of particles that is easy to note and qualitatively distinguish between different samples. Quantitative measurement of shape is much more difficult because of the three-dimensionality of particles and nonuniformity of size. Again, if we were living in a world of spherical particles—or, for that matter, particles of any regular geometric shape—then the parameter of shape would be easier to cope with.

Shape is of importance with respect to particles because it affects their movement in fluids and among themselves in the absence of a suspending fluid. A particle that has many irregular protrusions on its surface will have different static and dynamic behavior from a particle that is smooth and uniform. Irregular particles will undoubtedly stick together and behave as a single larger particle more readily than spherical or cubic particles. As the size of the particles becomes smaller and smaller, often the importance of shape becomes less. There is thus a relationship between size and shape.

The term *sphericity* is often applied to the measure of shape of a particle. Sphericity is defined as

$$\psi = \frac{\text{surface area of a sphere having the same volume as the particle}}{\text{surface area of the particle}} \tag{1-9}$$

The more the value of ψ deviates from 1.0 the less spherical the particle. See Example 1-5.

Often, particles have different profiles, depending on how they are setting on the solid surface upon which they are viewed. The projected length as seen by optical technique thus may vary. Hopefully, when many particles are viewed, on the average the particles will orient themselves on the solid surface to give a true average of the projected length. Statistical analysis indicates that at least 200 particles should be viewed to obtain an average length.

Using shape in a quantitative manner is most desirable. A series of microscopic photographs as given by McCrone and Delly (1973) may be used to compare the shape of a sample to samples that have been classified as to shape. Possibly, the flow behavior of a known particulate sample could be used to predict that of an unknown sample, assuming other physical and chemical processes do not come into play.

Classification of the shape of particles by the technique of Heywood (1963), as described by Allen (1975), seems most appropriate. Heywood uses two characteristics to describe shape. One characteristic is the degree of approaching a standard geometric shape and the other is that of relative proportions of the particle.

Consider Fig. 1-5 for definition of length (L'), width (W'), and height (H') of a particle. These measurements are combined into ratios termed *elongation* and *flakeness*.

$$\text{Elongation} = n' = \frac{L'}{W'}$$

$$\text{Flakeness} = m' = \frac{W'}{H'}$$

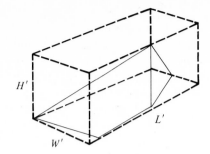

Figure 1-5 General particle diagram. (Heywood, (1963). Reproduced by permission of *J. Pharm. Pharmacol.*

The surface and volume of the particle in general can now be determined:

$$\text{Surface of particle} = S = fD_a^2 = \pi D_s^2 \tag{1-10}$$

$$\text{Volume of particle} = V = kD_a^3 = \pi D_v^3/6 \tag{1-11}$$

In these expressions D_a is the diameter of a circle having the same projected area as the particle in a stable position. The diameters D_s and D_v are the diameters of the particle based on the surface and volume averaging as shown in Eqs. 1-5 and 1-6. In the above equations, f and k are the surface and volume coefficients that Heywood determined experimentally. The expressions for f and k by empiricisms are

$$f = 1.57 + c\left(\frac{ke}{m'}\right)^{4/3}\left(\frac{n'+1}{n'}\right) \tag{1-12}$$

$$k = \frac{ke}{m'\sqrt{n'}} \tag{1-13}$$

The parameter ke is obtained for equidimensional particles and is given in Table 1-7 along with the parameter c. The breadth (m') to thickness (n') parameters are also incorporated in the expression. Figure 1-5 shows the dimensions of (n', m') for the given object. With these values the shapes of irregular particles may be approximated to give values of the surface area and volume of the particles.

Table 1-7 Shape coefficients for equidimensional particles (Heywood, 1963)

Shape group	ke	c
Geometric form		
tetrahedral	0.328	4.36
cubic	0.696	2.55
spherical	0.524	1.86
Approximate form		
angular, tetrahedral	0.38	3.3
prismoidal	0.47	3.0
subangular	0.51	2.6
rounded	0.54	2.1

Reproduced by permission of Heywood, H., *J. Pharm. Pharmacol. Suppl.* **15**:56T (1963).

Modern image analyzers permit fine scrutiny of individual particles such that a shape function can be analytically determined. Assuming that the group of particles analyzed have a random orientation on the microscope slide, a series of horizontal chords through the particle can be related to their distances from the center of the particle. This type of relationship can be utilized with statistical techniques of stereology to account for nonspherical shapes of particles in various analyses.

In addition the shape of particles can be analyzed by taking the profile of a particle and transforming this to a linear scale with the deviations around a reference value. The protrusions and indentations of the particle fluctuate about this reference level and Fourier Transforms are utilized to describe these variations. By this technique Fourier coefficients for various particle shapes can be catalogued and classified for reference.

Example 1-5 Compute the sphericity for the following common shapes: (*a*) cube (L'); (*b*) tetrahedron (L'); (*c*) rectangle parallelpiped (L', D', H').

(*a*) Cube:

$$\psi = \frac{\pi\,[(6/\pi)^{1/3}\,L']^{\,2}}{6L'^2} = 0.806$$

(*b*) Tetrahedron:

$$\psi = \frac{\pi\,[(6/\pi)^{2/3}(0.11785)^{2/3}\,L'^2]}{1.732L'^2} = 0.67$$

(*c*) Rectangle parallelepiped:

$$\psi = \frac{\pi\,[(6/\pi)(L'H'D)]^{\,2/3}}{2H'(2L'+D)} = \frac{\pi(6L'H'D)^{2/3}}{2H'(2L'+D)}$$

where $D = H' = \dfrac{L'}{2}$

$\psi = 0.767$

1-4 SIEVING TECHNIQUES

Dry Sieving

Sieving is by far the most common, easiest, and least expensive sizing technique available to the investigator. The American Society for Testing and Materials (ASTM, 1972) has prepared manuals on sieving methods in general and on techniques to handle many materials from coal to soap. The U.S. standard sieve series and the Tyler series are the two most common sieve series used in the United States. Tables 1-8 and 1-9 give the listing of these two series and their nominal sieve openings. There are four different specifications on sieves that can be considered. The wire cloth sieves are the most commonly used. The other sieves are the perforated plate, precision electro-formed, and matched sieves. The standard wire cloth sieves are usually made with brass frames and wire cloth. Stainless steel sieve cloth is finding wide acceptance due to its durability. The perforated sieve plates are available with either square or round apertures. A comparison of the results with a wire cloth sieve and the two types of perforated sieve plates indicates that the square apertures should be chosen. Precision electroformed sieves are available in sizes down to 5 μm. These sieves must be handled with great care because of the delicate electroformed sheet from which the sieves are made. Matched sieves have certain manufacturing tolerances, with a slight variation in average openings for each sieve. These sieves are tested with respect to a standard sieve or master set.

Table 1-8 U.S. sieve series (Perry, 1965)

Meshes linear in	Sieve No.	Sieve opening, in	Sieve opening, mm	Wire diameter, in	Wire diameter, mm
2.58	$2\frac{1}{2}$	0.315	8.00	0.073	1.85
3.03	3	0.265	6.73	0.065	1.65
3.57	$3\frac{1}{2}$	0.223	5.66	0.057	1.45
4.22	4	0.187	4.76	0.050	1.27
4.98	5	0.157	4.00	0.044	1.12
5.81	6	0.132	3.36	0.040	1.02
6.80	7	0.111	2.83	0.036	0.92
7.89	8	0.0937	2.38	0.0331	0.84
9.21	10	0.0787	2.00	0.0299	0.76
10.72	12	0.0661	1.68	0.0272	0.69
12.58	14	0.0555	1.41	0.0240	0.61
14.66	16	0.0469	1.19	0.0213	0.54
17.15	18	0.0394	1.00	0.0189	0.48
20.16	20	0.0331	0.84	0.0165	0.42
23.47	25	0.0280	0.71	0.0146	0.37
27.62	30	0.0232	0.59	0.0130	0.33
32.15	35	0.0197	0.50	0.0114	0.29
38.02	40	0.0165	0.42	0.0098	0.25
44.44	45	0.0138	0.35	0.0098	0.22
52.36	50	0.0117	0.297	0.0074	0.188
61.93	60	0.0098	0.250	0.0064	0.162
72.46	70	0.0083	0.210	0.0055	0.140
85.47	80	0.0070	0.177	0.0047	0.119
101.01	100	0.0059	0.149	0.0040	0.102
120.48	120	0.0049	0.125	0.0034	0.086
142.86	140	0.0041	0.105	0.0029	0.074
166.67	170	0.0035	0.088	0.0025	0.063
200	200	0.0029	0.074	0.0021	0.053
238.10	230	0.0024	0.062	0.0018	0.046
270.26	270	0.0021	0.053	0.0016	0.041
323	325	0.0017	0.044	0.0014	0.036

Table 1-9 Comparison of Tyler and U.S. sieve series (Perry, 1965)

Tyler standard sieve series				U.S. series
Opening, in	Opening, mm	Tyler mesh	Diameter of wire, in	approximate equivalent no.
3	......	......	0.207	
2	......	......	0.192	
$1\frac{1}{2}$	......	......	0.148	
1.050	26.67	......	0.148	
0.883	22.43	......	0.135	
0.742	18.85	......	0.135	
0.624	15.85	......	0.120	
0.525	13.33	......	0.105	
0.441	11.20	......	0.105	
0.371	9.423	......	0.092	
0.312	7.925	$2\frac{1}{2}$	0.088	
0.263	6.680	3	0.070	
0.221	5.613	$3\frac{1}{2}$	0.065	
0.185	4.699	4	0.065	4
0.156	3.962	5	0.044	5
0.131	3.327	6	0.036	6
0.110	2.794	7	0.0328	7
0.093	2.362	8	0.032	8
0.078	1.981	9	0.033	10
0.065	1.651	10	0.035	12
0.055	1.397	12	0.028	14
0.046	1.168	14	0.025	16
0.0390	0.991	16	0.0235	18
0.0328	0.833	20	0.0172	20
0.0276	0.701	24	0.0141	25
0.0232	0.589	28	0.0125	30
0.0195	0.495	32	0.0118	35
0.0164	0.417	35	0.0122	40
0.0138	0.351	42	0.0100	45
0.0116	0.295	48	0.0092	50
0.0097	0.0246	60	0.0070	60
0.0082	0.208	65	0.0072	70
0.0069	0.0175	80	0.0056	80
0.0058	0.0147	100	0.0042	100
0.0049	0.0124	115	0.0038	120
0.0041	0.0104	150	0.0026	140
0.0035	0.089	170	0.0024	170
0.0029	0.074	200	0.0021	200
0.0024	0.061	250	0.0016	230
0.0021	0.053	270	0.0016	270
0.0017	0.043	325	0.0014	325
0.0015	0.038	400	0.0010	

Reproduced by permission of McGraw-Hill Book Company, *Chemical Engineers Handbook.*

The degree to which particles are divided into their respective size classifications in the sieving process depends on a number of parameters. The particle size distribution is important—whether narrow, broad, bimodal, etc. This size distribution will set the amount of material on a particular sieve—whether sparse or overloaded. The physical properties of the particles are in many respects the controlling factors. Moisture, degree of agglomeration, shape, electrostatics, and stickiness all come into play. Once the particles are placed on the top sieve, the method of agitating the sieve set is of concern. The motion must not always be in one direction but in many directions to ensure proper sieving of irregularly shaped particles. Various mechanical and sonic sifters are on the market to provide the proper degree of motion for sieving. In addition to the proper motion, the proper duration of sieving is necessary. The sieve itself must be in good condition, guaranteeing a uniform set of openings over the cross section. As time goes on, the sieve itself will wear through use, causing larger openings in the sieve than the original design. This must be carefully monitored. Proper cleaning after each test is imperative in order to ensure reproducibility on repeated sieving. Proper weighing with accurate balances is another factor that may profoundly affect the result of sieving. Example 1-6 shows an actual sieve analysis on a specific coal.

Wet Sieving

Particles often have an adhering or agglomerating property, mainly due to their moisture content. Because of this adhering property, the particles are difficult to sieve using the normal dry method. In these cases, water is generally fed into the top sieve with the solid sample and rinsing and vibration occur simultaneously. The sieve and sample are then dried and weighed. It is suggested by the ASTM not to exceed 110°C in the drying process of these sieves. It should be remembered that in sizing particles in this way the sizing may not be representative of the true particle size experienced in the actual process.

Example 1-6

Sieve analyses of a Montana rosebud coal

Mesh size	Tare, g	Weight, g	Δ Weight, g	%
20	45.0	45.1	0.1	1
40	40.1	40.6	0.5	5.1
80	36.2	38.3	2.1	21.2
100	35.5	36.6	1.1	11.1
140	35.0	36.4	1.4	14.1
200	33.4	34.8	1.4	14.1
Pan	160.7	164.0	3.3	33.3
				99.9

Calculation of weight average diameter from sieve analyses

Mesh size	Mesh opening, μm	%	Mesh opening × %
20	840	1	840
40	420	5.1	2,142
80	177	21.2	3,735
100	149	11.1	1,654
140	105	14.1	1,481
200	74	14.1	1,043
Pan	37	33.3	1,232
		99.9	12,127

$$\frac{\Sigma \text{ mesh opening} \times \%}{99.9} \approx 121 \ \mu\text{m weight average diameter}$$

1-5 MICROSCOPE SIZING TECHNIQUES

After sieving, optical microscopic techniques are probably the most common methods used for particle sizing. Not only are sizes obtained, but shapes can be discriminated. The power of microscopic analysis has increased dramatically in the last several years. Sizing can proceed down to angstrom levels with the use of the electron microscope. Recent advances in counting, recording, and analysis with electronic and computer adaptations have changed a once very tedious, blinding experience into a simple operation. It is now possible to size thousands of particles by the flick of a finger. At present, cost is the only factor that prevents widespread use of these modern advances.

Optical Microscope

In order to analyze the particles under microscopic examination they must be placed on a surface for viewing. A glass microscopic surface is most appropriate for optical microscopes. The major task in preparation of a good slide is a proper dispersion of the particles so they can be sized without interference from neighboring particles. For producing permanent slides where the dispersant liquid serves as an adhesive on drying, Orr and Dallavalle (1960) suggest placing a small amount of particulate matter in a 10^{-5} m^3 beaker and adding 2 or 3 × 10^{-6} m^3 of 2% colloidon in butyl acetate. A drop of this is then placed on the surface of still water, forming a film that can be picked up on a glass microscopic slide. Other techniques that are easier to handle involve placing the suspension directly on the glass slide and letting the solvent evaporate directly, or placing a cover slide on the droplet and sliding this over the base slide to achieve further dispersion. Other dispersing solutions with adhesive properties may also be used—for example, colloidon in amyl acetate, Canada balsam in xylol, and polystyrene in xylene.

If nonpermanent slides are acceptable, a simple combination of dispersant liquid and the solid material can be placed on a slide. A slip cover, as mentioned above, may be used or a small camel hair brush may be used for dispersing the liquid droplet on the slide. In some cases, use of a dispersant liquid may not be desirable and simple sprinkling of dry particles on a slide may be preferred, depending on the end process in which the particles will be found.

Electron Microscope

Two types of electron microscopy are available for analyzing particles. The transmission electron microscope (TEM) can look at particles down to 10 to 20 Å. This operation produces an image on a fluorescent screen or photographic plate by means of an electric beam. See Fig. 1-6 of Example 1-7 for an example of a TEM photograph. The scanning electron microscope method (SEM) uses a fine beam of electrons to scan across a sample in a series of parallel tracks. These electrons interact with the sample, producing a secondary back scattering or cathode luminescence and x-rays. A resolution of about 200 Å can be achieved. Details of these electron microscopic techniques are available in numerous books (Allen, 1975; McCrone and Delly, 1973). See Fig. 1-7 of Example 1-7 for an example of a SEM photograph.

Graticules

In manual analysis of individual or multiple particles, the stage graticule is often used. The graticule is a microscope slide that has an engraved linear scale or circles of prescribed sizes on its surface for making comparisons with the samples being viewed. See Fig. 1-8 in Example 1-8.

Image Analyzers

The easiest method of analyzing particles by optical microscopic techniques is by use of quantitative image analyzers. The Bausch & Lomb Omnicon Image Analyzer and Quantimet are two such instruments. These automated instruments make it easy to size the 600+ particles that are generally recommended for analyses to give a statistically accurate count. Multiple fields of view are also possible with these instruments.

Example 1-7 Photographs of particles using an electron microscope are seen in Figs. 1-6 and 1-7.

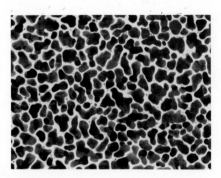

Figure 1-6 A 200-Å tin oxide discontinuous film with TEM. (Photograph by H. Moyes, University of Pittsburgh.)

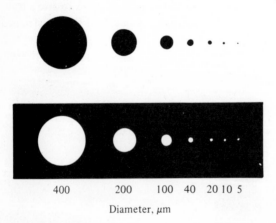

Figure 1-7 SEM of pulverized Pittsburgh seam coal, 24° tilt. (Photographs by J. Blachere, University of Pittsburgh.)

Example 1-8 An example of a graticule for particle sizing is seen in Fig. 1-8.

400   200   100 40   20 10 5

Diameter, μm

Figure 1-8 Graticule. (Reproduced by permission of Bausch & Lomb.)

OTHER TECHNIQUES FOR SIZE DETERMINATION

Sedimentation

The sedimentation technique is based on the ability of the gravity force to distinguish among different sizes of particles. A homogeneous mixture of solids and fluid can be studied, or a thin layer of particles can be introduced at the top of the fluid to watch its settling progress. An incremental analysis will give the change in concentration with time at a fixed position, and the cumulative analyses will determine the rate of settling. From this information, a size distribution can be found. In the pipette method of sedimentation, studies are made of samples of a definite volume to note concentration changes and size distributions. The Andreasen pipette method has seen wide application. In this technique, the pipette itself remains in the apparatus. Figure 1-9 shows a typical example of a size distribution using the Andreasen pipette technique.

The sedimentation process can also be monitored by a photosensor composed of a narrow beam of light projecting through a settling mixture. The degree of light attenuation is in direct correspondence to the solids concentration. This technique does not work well for opaque liquids, and acoustic transmitters have recently been used in applications on opaque liquids. The absorption of the acoustic signal is measured by the transducer and this can be correlated with the solids concentration. In addition to using light and sound energies to monitor the settling progress, x-rays have been applied successfully in the x-ray sedimentation units. The absorption of the

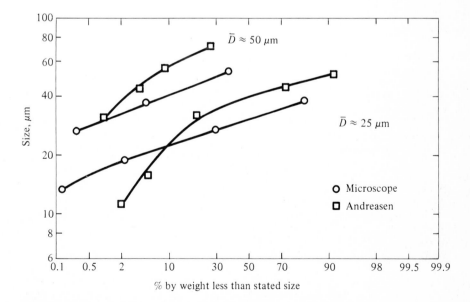

Figure 1-9 Comparison of weight distributions by optical microscopy and Andreasen pipette methods (Peters, 1971). Reproduced by permission.

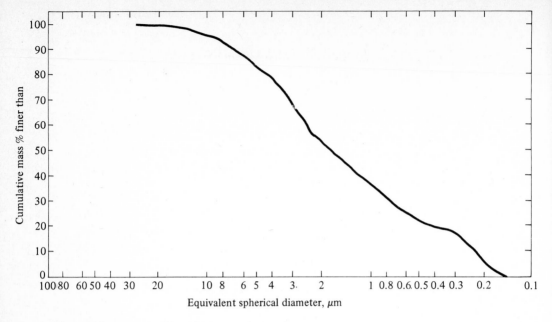

Figure 1-10 An x-ray sedimentation curve for urania.

x-ray is dependent on Beer's law of absorption. A number of commercial x-ray sedimentation units are on the market. Figure 1-10 shows the results of an x-ray sedimentation analysis performed on urania.

When the sedimentation process due to gravity is too slow for measurement of particle size, centrifugal forces can be imposed on the particles up to speeds of 100,000 rpm in an ultracentrifuge. The same light, acoustic, and x-ray techniques can be applied to centrifugal methods to note change in concentration and thus the particle size distribution.

Coulter Counter

The Coulter counter is an instrument that is now being used more and more in the determination of particle size distributions. The particles must be suspended in an electrolyte that is passed through a small orifice. The orifice has electrodes on either side that generate voltage pulses with amplitudes proportional to the volume of the particles of known diameter in known electrolytes. The Coulter counter gives good results in sizing down to 1 μm-diameter particles, although smaller sizes can be handled in the instrument. Table 1-10 gives an example of a Coulter counter analysis on latex beads.

Light Scattering

Light scattering can also be employed as a means of sizing particles. Particles of different sizes and concentration will scatter light in various directions. A photomultiplier tube generally picks up the scattering light. The signal obtained is proportional to the size and concentration of the particles. Generally, this technique of sizing has been

Table 1-10 Coulter counter analysis for 2.02-μm latex beads on an industrial model B Coulter counter (Miller, 1971)

Lower threshold, t_1	Aperture current, i	Amplifier section, A	Particle volume, $V = kt_1 iA$	Particle diameter, D	n	Δn	$\bar{V}$	Weight fraction
20	1	32	861.4	11.80	4	6	646	0.029
20	1	16	430.7	9.43	10	15	323	0.065
20	1	8	215.4	7.45	25	43	1,615	0.117
20	1	4	107.7	5.88	68	65	80.8	0.156
20	1	2	53.84	4.68	133	21	47.1	0.164
15	0.5	4	40.38	4.25	112	19	39.22	0.219
20	0.707	2	38.06	4.17	131	15	333	0.223
15	0.354	4	28.59	3.80	146	492	27.8	0.325
20	1	1	26.92	3.72	638	736	20.2	0.436
20	1	0.5	13.46	2.95	1,374	2,192	10.1	0.602
20	1	0.25	6.73	2.34	3,566	2,356	5.74	0.702
20	0.707	0.25	4.76	2.08	5,922	6,363	4.06	0.896
20	0.5	0.25	3.36	1.86	12,285	4,842	2.87	1.00
20	0.354	0.25	2.38	1.65	17,127	–	–	–

Note: Calibration factor $k = 1.346$, aperture resistance $= 60,000 \ \Omega$, gain control $= 100$, Matching switch $= 16\text{-}H$, aperture diameter $= 50 \ \mu$m, electrolyte is 1% NaCl.

employed on very small particles (10 μm or less). The laser is ideally suited as a light source for this method. In the field of aerosol physics, the light-scattering measuring devices are frequently used. Royco Instruments, Inc., and Bausch & Lomb manufacture light-scattering devices for measuring particle size distributions of aerosols.

1-7 MIXING

The handling of two or more different types of solids often requires mixing. The task of solid mixing is not as easy as it sounds, especially if one has had previous experience only in the liquid mixing area. Often, when two powders are mixed, a high order of mixing can be achieved. However, when the mix is then moved from the blender to the process, a degree of segregation occurs, which undoes a certain amount of the blender's action. This segregation could be significant if tight tolerances are set on the final product. The ideal situation would be to have an in situ mixing process for direct utilization of the solids.

There are many mixers on the market that can serve a number of purposes in the solid mixing area. Table 1-11 lists some of these mixers and a few of their characteristics. Williams (1963, 1968, Williams and Khan, 1973) has written extensively on solid powder mixing and has considered all aspects of the process. There are certain guidelines one can follow in this field, but the area is far from an exact science. A very comprehensive compilation of mixing literature has been given by Cooke, Stephens, and Bridgewater (1976).

Table 1-11 Mixer characteristics (Williams, 1968)

Type of mixer	Main mixing mechanism	Segregation
Horizontal drum	Diffusive	Sizable
Slightly inclined drum	Diffusive	Moderate
Lödige mixer	Convective	Little
Stirred vertical cylinder	Shear	Sizable
V mixer	Diffusive	Sizable
Y mixer	Diffusive	Sizable
Double cone	Diffusive	Sizable
Ribbon blender	Convective	Little
Fluidized mixer	Convective	Moderate
Nauta mixer	Convective	Little

In quantifying the goodness of a mix the standard deviation is often employed. This is generally sufficient for a powder that is normally distributed in composition. Imperfect mixing gives distributions that are far from the normal. The standard deviation can be used to classify the mix, but care must be taken when the proportion of samples is outside a specified range where nonnormal composition distributions exist.

Certain specifications must be met in analyzing the state of mixing. When referring to a component, one must know the sample size, the limits on the percentage of the component present in the sample, and the frequency the samples must lie between. Generally, a large number of samples should be taken to analyze the mix. Valentin (1967) recommends that at least 50 samples be taken to give a reliable measure of the standard deviation. Fewer than 20 samples give unreliable results.

When performing a mixing operation, the mixer type is of paramount importance. The various mixers can produce varying degrees of segregation (see Table 1-11). In general, segregation can be caused by pouring the solids into a heap, by vibrations, or by stirring. The mixing mechanisms are convective, diffusive, or shear in nature. Generally, one type of mechanism dominates in a particular mixer.

If one has a dry powder that is showing segregation, addition of a small amount of water often diminishes this tendency for segregation. Cohesive powders also are seen to have low segregation ability. The causes of segregation are the differences in the properties of the components, especially the size distribution. Other factors of importance are the density, shape, roughness, and resilience of the particles. In general, free-flowing materials will tend to segregate. Figure 1-11 shows the effect of diameter ratios of components on segregation as measured by the standard deviation.

There is a particular phenomenon that occurs with fine powders, and that is their tendency to form what is known as soft agglomerates. These agglomerates have weak binding forces and will disintegrate on touching; however, they present troublesome obstacles to forming homogeneous mixtures of fine materials. In the mixing operation the agglomerates can maintain their integrity and be of sufficient size to cause local inhomogeneity in the mixture.

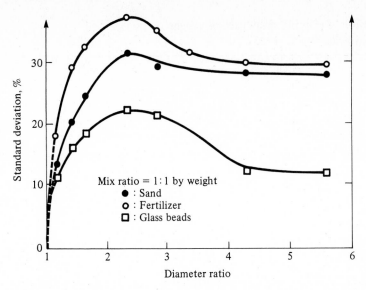

Figure 1-11 Variation of standard deviation with diameter ratio of components for V mixer II (Williams and Khan, 1973). Reproduced by permission of the Institution of Chemical Engineering.

There are many ways to characterize powders but only the method of Poole (Poole, Taylor, and Wall, 1964; Poole, 1965) will be presented, because recent experience has proved it to be adequate. This technique is based on a single number for characterization of the powder, G, the mean particle weight, which is defined as

$$G = \Sigma\, fw \qquad (1\text{-}14)$$

where
f = weight fraction in each class
w = individual particle weight for each class

Poole has shown typical calculations to determine the mean particle weight for urania powder, as given in Table 1-12.

The values of G for various powders or agglomerates can be used when mixing occurs to classify the goodness of a mix. This calculation employs Stange's (1954) two powders approximation to the theoretical standard deviation of the sample concentration in a fully randomized mix, σ_R:

$$\sigma_R^2 = \frac{xy}{M}(xG_y + yG_x) \qquad (1\text{-}15)$$

where
M = sample weight
x = weight fraction of component x
y = weight fraction of component y

Table 1-12 Calculation of effective mean particle weight for a typical urania powder (Poole, 1965)

Diameter of particle, μm	Particle weight, μg	Mean particle weight of range, μg	Fraction under size, f'	Range fraction, f	fw, μg
2,411	36,719		1.00		
		> 29,761		> 0.013	387
2,057	22,803		0.987		
		> 17,569		> 0.019	334
1,676	12,334		0.968		
		> 9,801		> 0.025	245
1,405	7,267		0.943		
		> 5,920		> 0.034	201
1,204	4,573		0.909		
		> 3,603		> 0.047	170
1,003	2,643		0.862		
		> 1,769		> 0.084	149
699	895		0.778		
		> 612		> 0.133	81
500	328		0.645		
		> 185		> 0.379	70
251	41		0.266		
		> 25		> 0.124	3
152	9		0.142		
		> 5		> 0.127	1
53	0.3		0.015		
		> 0.15		> 0.015	0
0	0		0		

$$G = \Sigma\ fw = 1{,}640\ \mu g$$

Reproduced by permission of British Ceramics Society.

For some free-flowing sample materials, Poole has considered an example where the values of G are given as follows:

Substance	G, μg
Copper	0.821
Nickel	2.206
Urania	0.186
Urania (rod-milled)	0.076
Thoria	0.0148
Plutonia	0.387

Poole has analyzed various mixes of these materials with σ_R from Stange's formula and the standard deviation. The standard deviation was calculated at a certain point in the mixing process after which no further reduction occurred, and s was compared to σ_R by a ratio of the two values. Some values of σ_R are given below:

Mix material	Ratio	σ_R	s	s/σ_R
Copper and nickel	110:1	21.9×10^{-5}	25.4×10^{-5}	1.16
	1000:1	7.19×10^{-5}	7.80×10^{-5}	1.09

The closer the value of the ratio of s/σ_R to 1.0, the better the mixing.

Thus, the value of the mean particle weight of a powder can be employed as a characterizing parameter of a powder, and then this may be incorporated in the measure of goodness of a mix. Example 1-9 shows an application of this measure of goodness of a mix.

Example 1-9 In the mixing of two ceramic materials the concentration fraction of the minor component is 0.0215 and that of the major component is 0.9785. The effective mean particle size of the minor species is 3.28×10^{-9} μg and that of the major species 6.55×10^{-9} μg. The sample mass is 0.5×10^6 μg.

Using this information one can calculate the theoretical standard deviation of the sample concentration of a freely randomized mix as

$$\sigma_R^2 = \frac{xy}{M}(xG_y + yG_x)$$

$$\sigma_R^2 = \frac{(0.0215)(0.9785)}{0.5 \times 10^5}[0.0215(6.55 \times 10^{-9}) + 0.9785(3.28 \times 10^{-9})]$$

$$\sigma_R^2 = 1.41 \times 10^{-10}$$

$$\sigma_R = 1.19 \times 10^{-5}$$

In the mixing process the standard deviation was determined as a function of time as follows:

Mixing time, sec	s	s/σ_R
1,000	0.432	0.363×10^5
2,000	0.124	0.104×10^5
4,000	0.027	0.023×10^5
8,000	0.020	0.017×10^5
12,000	0.034	0.029×10^5
24,000	0.020	0.017×10^5

The ratio s/σ_R decreases with time. The variation of s/σ_R from unity indicates agglomeration tendencies of these materials.

PROBLEMS

1-1 The table gives a size distribution for a Pittsburgh seam coal injected directly into a combustion. What particle diameter should be used in analyzing the combustion process and what is its value?

Cumulative number of particles for microscope field of view

Diameter, μm	1	2	3	4
1	1,492	1,321	1,617	1,665
2	1,242	1,168	1,329	1,284
3	836	876	892	872
4	690	706	710	589
5	503	494	474	420
6	388	379	377	328
7	335	335	318	295
8	278	307	286	233
9	253	280	256	209
10	209	235	222	182
15	124	139	102	78
20	70	89	53	38
25	38	50	29	20
40	14	20	4	7
50	7	12	1	3
75	3	2	0	1

1-2 The ceramic powder described in the table has been found to flow with considerable facility in loading and unloading operations in a bin. Can you analyze the powder distribution to find any reasons for this behavior?

Diameter, μm	Cumulative mass % finer than	Diameter, μm	Cumulative mass % finer than
75	100	22	53
70	99	21	51
65	97	20	50
60	95	19	38
55	92	18	36
50	80	17	33
45	75	16	30
40	72	15	25
35	69	14	20
30	65	13	15
25	62	12	10
24	60	11	5
23	57	10	0

1-3 Compute the particle statistics on the two particle samples from the same process given in the table. Note similarities and differences. What conclusion can you make about these two samples?

	Number of particles	
Diameter, μm	Sample 1	Sample 2
1	1,353	248
5	485	100
10	234	43
20	85	8
25	56	5
40	20	2
50	17	2
75	4	1

1-4 Two irregular particles have the following profile dimensions:

Particle 1:	$H' = 50\ \mu$m	Particle 2: $H' = 100\ \mu$m
	$W' = 60\ \mu$m	$W' = 200\ \mu$m
	$L' = 100\ \mu$m	$L' = 300\ \mu$m

Particle 1 is approximately tetrahedral and particle 2 is subangular. Determine the surface areas and volumes of the particles.

1-5 Determine the length, volume-surface, and weight diameters for the following sizing of Montana rosebud coal:

Size, μm	Cumulative number of particles	Size, μm	Cumulative number of particles
1	1,249	40	40
5	480	50	15
10	216	75	8
15	114	100	3
20	78	150	2
25	47	200	1

1-6 Determine the weight average diameter from the sieve analysis on the following Montana rosebud coal:

Mesh size	Tare, g	Weight after sieving, g
20	44.8	44.9
40	39.9	40.6
80	35.9	40.1
100	35.2	36.4
140	35.0	35.9
200	33.3	34.0
Pan	156.9	159.1

REFERENCES

Allen, T.: *Particle Size Measurement,* Chapman and Hall (Wiley), New York, 1975.

American Socity for Testing and Materials: *Test Sieving Methods,* ASTM, Philadelphia, 1972.

Cooke, M. H., D. J. Stephens, and J. Bridgewater: *Powder Technol.* **15**:1 (1976).

Heywood, H.: *J. Pharm. Pharmacol. Suppl.* **15**:56T (1963).

McCrone, W. C., and J. G. Delly: *The Particle Atlas,* Ann Arbor Science Publishers, Ann Arbor, Mich. 1973.

Miller, E. M. S.: Thesis, *The Effect of Bed Material Size on Aerosol Collection, Using A Fluidized Bed,* University of Pittsburgh, 1975.

Orr, C., and J. M. Dallavalle: *Fine Particle Measurements,* Macmillan, New York, 1960.

Perry, J. H. (ed.): *Chemical Engineers' Handbook,* McGraw-Hill, New York, 1965.

Peters, L. K.: Ph.D. Thesis, *A Study of Two-Phase Solid in Air Flow Through Rigid and Compliant Wall Tube,* University of Pittsburgh, 1971.

Poole, K. R.: *Brit. Ceramic Soc. Proc.* **3**:43 (1965).

——, R. F. Taylor, and G. P. Wall: *Trans. Inst. Chem. Eng.* **42**:T305 (1964).

Stange, K.: *Chem. Eng. Tech.* **26**:331 (1954).

Valentin, F..H. H.: *Trans. Inst. Chem. Eng.* **45**:CE99 (1967).

Williams, J. C.: *Univ. Sheffield Fuel Soc. J.* **14**:29 (1963).

——: *Powder Technol.* **2**:13 (1968).

—— and M. I. Khan: *Chem. Eng.* **209**:19 (Jan. 1973).

ADHESION AND AGGLOMERATION

2-1 INTRODUCTION

Adhesion and agglomeration are vitally important in the gas-solid transport area although they have not received much attention in that respect. There have been numerous studies on these subjects relating the solid particles and surfaces to static conditions. As particles flow in a pipe, a certain number of the particles adhere to the solid surface while others find their way into small crevices in the system. In the flow of a fine powder in a pipe, after the initial deposition of the particles to the pipe wall, the particles in suspension will interact with those adhering to the wall, causing different frictional and electrostatic behavior than would be anticipated from interactions of the particles with the pipe wall material alone, whether it is metal, plastic, or glass.

It should be noted that the size of the adhesion force is indeed large on small particles. In many cases this force is several orders of magnitude greater than the gravity force on the particle. Because of these sizable adhesion forces, large forces are needed to remove the small particles from the adhering surfaces.

Agglomeration of solid particles comes about from a myriad of conditions: the types of particle-particle interactions, the surface character of the solid, moisture, particle and surface stickiness, surface roughness, particle size, etc. Knowledge of the condition of the solids in the actual flow system is of vital importance. Our static methods of sizing and analyzing particles often bear no resemblance to the actual state present in a flow system of solids in a gas stream. The effective average flow diameter differs markedly from the static measurements. This fact is one of the major reasons why there are some wide discrepancies among investigators on the exact behavior of gas-solid types of flow systems.

2-2 ADHESION THEORIES AND EFFECTS

In considering adhesion forces the van der Waals and electrostatic forces are considered as the dominant long-lasting factors. Other effects are capillary forces, chemical bonding, sintering, and alloying (which may be important in certain cases). The calculation of the van der Waals forces is based on the macroscopic theory of Lifshitz, which is coupled with the imaginary parts of the complex frequency-dependent dielectric constants of the adherents. Potential theory and boundary-layer theory of semiconductors can be used to estimate the action of the electrostatic force. A quantitative appraisal of this theory is not yet possible. However, the theory is able to predict order of magnitude values that are in agreement with experimental findings.

Van der Waals Forces

The method considered here for determination of these forces is the modified Lifshitz macroscopic approach, considering the microscopic approach results. This approach uses the optical properties of interacting macroscopic bodies to calculate the van der Waals attraction from the imaginary part of the complex dielectric constants. The other possible approach—the microscopic theory—uses the interactions between individual atoms and molecules postulating their additive property. The microscopic approach, which is limited to only a few pairs of atoms or molecules, has problems with the condensation to solids and ignores the charge-carrier motion. The macroscopic approach has great mathematical difficulties, so the interaction between two half-spaces is the only one to be calculated (Krupp, 1967).

As a model consider two solids, as shown in Fig. 2-1. The solids are separated from each other by a gap of width Z_0. The internal electromagnetic fluctuation fields in solids set up an electromagnetic field E_3, H_3 in the gap. The resultant electromagnetic field strengths E_3 and H_3 of the gap are used to determine the Maxwell electromagnetic stress tensor T. The force component parallel to the Z axis is

$$\frac{\partial^2 \overline{T}}{\partial Z^2} = \frac{1}{2} \left\{ \epsilon_0 \left[\left(\frac{\partial E}{\partial Z}\right)^2 - \left(\frac{\partial E}{\partial X}\right)^2 - \left(\frac{\partial E}{\partial Y}\right)^2 \right] + \mu_0 \left[\left(\frac{\partial H}{\partial Z}\right)^2 - \left(\frac{\partial H}{\partial X}\right)^2 - \left(\frac{\partial H}{\partial Y}\right)^2 \right] \right\} \quad (2\text{-}1)$$

where ϵ_0 = dielectric constant
μ_0 = magnetic permeability

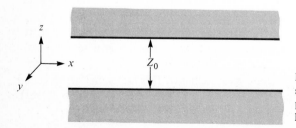

Figure 2-1 Distance between two solids (Krupp, 1967). Reproduced by permission of Elsevier Scientific Publishing Co.

After some mathematical effort, Lifshitz found the van der Waals adhesion pressure to be

$$P_{VDW} = \frac{\hbar\omega}{8\pi^2 Z_0^3} \tag{2-2}$$

where a frequency ω is

$$\omega \equiv \int_0^\infty \frac{\epsilon_1(i\xi) - 1}{\epsilon_1(i\xi) + 1} \frac{\epsilon_2(i\xi) - 1}{\epsilon_2(i\xi) + 1} d\xi \tag{2-3}$$

$\hbar$ = Planck's constant
Z_0 = adhesional distance ≈ 4 Å
ϵ_i = dielectric constant
ξ = imaginary frequency

The imaginary part of the dielectric constant is a measure of the energy dissipation into heat that an electromagnetic wave of angular frequency ω suffers in a given medium.

Each solid body contains local electric fields of a wide range of frequencies. These fields originate from spontaneous polarization of the atoms or molecules. The intensity of the fields increases with increasing optical absorptivity, i.e., with an increasing imaginary part of the dielectric constants. These field correlations, or mutual dragging into phase, are at the origin of the van der Waals forces.

Using the microscopic theory, the pressure between a sphere and a half-space has been shown by Hamaker (1937) to be

$$P_{VDW} = \frac{A'}{6\pi Z_0^3} \tag{2-4}$$

Considering the area of contact in the sphere and half-space geometry, Hamaker calculated the force of adhesion as

$$F_{VDW} = \frac{A'R}{6Z_0^2} \tag{2-5}$$

where R = radius of the particle
A' = Hamaker constant

Using Eq. 2-2 from the macroscopic theory, the force between a sphere and a half-space can be determined:

$$F_{VDW} = \frac{\hbar\omega R}{8\pi Z_0^2} \tag{2-6}$$

Corn (1966) has compiled a table of values of Hamaker constants for various adhesion processes. These Hamaker constants are seen to vary from 1×10^{-21} to 2×10^{-17} J.

Table 2-1 Van de Waals forces of attraction F_{VDW} (in 10^{-8} N) between a sphere (radius R) and a half-space at an adhesional distance of 4 Å between the adherents, as a function of Lifshitz–van der Waals constant $\hbar\omega$ (Krupp, 1967)

$\hbar\omega$, J	R		
	1 μm	10 μm	100 μm
0.96×10^{-19}	2	24	240
3.20×10^{-19}	8	80	800
$14.4 \ \times 10^{-19}$	36	360	3,600

Reproduced by permission of Elsevier Scientific Publishing Co.

Krupp (1967) has varied the value of $\hbar\omega$ for adherents of various sizes to calculate the value of the adhesion force. Table 2-1 presents the magnitude of these forces as a function of the particle size and $\hbar\omega$ constants. A calculation of the magnitude of the Hamaker constant and Lifshitz constant is seen in Example 2-1.

Example 2-1 Experimental determination of the adhesion force of alumina powder to a Pyrex glass surface has been carried out by a centrifuge method. A force of 2.25×10^{-5} N has been found to be necessary to remove a 17.5-μm-size particle. What is the value of the average frequency constant $\hbar\omega$ and the Hamaker constant for this system?

For $Z_0 = 4$ Å

$$\text{Hamaker:} \quad F_{VDW} = \frac{A'R}{6Z_0^2} = 2.25 \times 10^{-5} \text{ N} = A' \frac{(17.5/2) \times 10^{-6} \text{ m}}{6(4 \times 10^{-10})^2 \text{ m}^2}$$

$$A' = 24.7 \times 10^{-19} \text{ J}$$

$$\text{Lifshitz:} \quad F_{VDW} = \frac{\hbar\omega R}{8\pi Z_0^2} = 2.25 \times 10^{-5} \text{ N} = \frac{\hbar\omega(17.5/2 \times 10^{-6}) \text{ m}}{8\pi(4 \times 10^{-10} \text{ m})^2}$$

$$\hbar\omega = 101 \times 10^{-19} \text{ J}$$

$$\hbar = 6.63 \times 10^{-34} \text{ J-sec}$$

$$\omega = 15.2 \times 10^{15} \text{ sec}^{-1}$$

Electrostatic Forces

Basic principles of electrical forces on contact Electrostatic forces involved in the adhesion phenomena have the same basis as those forces involved in the static electrification process discussed in Chap. Six. The discussion presented here for adhesion will also serve as a base for the static electrification fundamentals.

Loeb (1958) has given a rather comprehensive analysis of electrification of metals by using the description of the movement of electrons and the work function related to the contact potential. In metals there is a segregation of electrons from the outer valence states of free atoms to the energy bands of electrons capable of moving relatively freely in the space between the ions of the metal lattice in regions of constant potential. These electrons fill the band to a fixed level. Because the band is partly empty or overlapping, there are lower negative energies below the zero energy at the states. As the temperature increases, these states are gradually filled. There is a potential field at the surface to prevent electrons from escaping. This potential field exists between the free electrons at the top of the Fermi band and the outside. This is called the intrinsic potential ϕ of the metal. The work function is given by the product of ϕ and the electronic charge e. The intrinsic potential is changed by a number of effects:*

$$
\begin{array}{ll}
\text{Increased temperature} & \phi\uparrow \\
\text{External field} & \phi\downarrow \\
\text{Adsorbed negative ions} & \phi\uparrow
\end{array}
$$

The contact potential between two metals 1 and 2 is represented in terms of the intrinsic potentials:

$$V_c = \phi_1 - \phi_2$$

For contacting between a metal and an insulator having few free electrons, contact electrification does not necessarily involve the loss or gain of electrons by the metal. The contact electrification between a nonmetal and a metal is most likely caused by electron transfer from the metal and ion transfer of either sign from the nonmetal. The electrons of the nonmetal surface or ions and the metal surfaces must be strong enough to overcome surface, image, and binding forces. Static electrification in nonmetal-metal systems is likely to be very effective due to the fact that charge cannot readily leak away on separation. The amount of charge is proportional to the effective areas of the surfaces in contact. Impacts of solids on the surface of the pipe can actually abrade the oxide films on metal surfaces and can cause sizable changes in contacting potentials.

As noted, the work required to remove an electron at the Fermi level to a point in free space just outside the solid is called the work function W_ϕ of the solid. In semiconductors use is made of an additional quantity, electron affinity χ, which is the work required to remove an electron from the bottom edge of the conduction band at the surface to a point in free space just outside the semiconductor. For the case of a semiconductor, all its bands are filled except one or two bands, which are slightly filled or slightly empty (Kittel, 1971).

In Fig. 2-2, for the semiconductor, E_c is the bottom edge of the conduction band and E_v is the top edge of the valence band. The quantity W_b is the energy difference

---

*Loeb (1958) Reproduced by permission of Springer-Verlag.

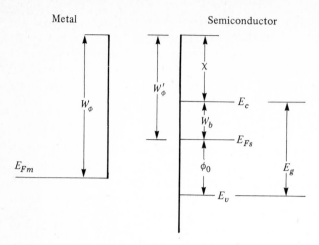

Figure 2-2 Noncontacted energy bands between a metal and a semiconductor (Krupp, 1971). Reproduced by permission of the Institute of Physics.

between the Fermi level E_{F_s} and the conduction band edge E_c with E_g as the band gap in the solid. The Fermi level of the metal is noted as E_{FM} with W_ϕ being its associated work function.

If these two materials (metal and semiconductor) are contacted, electrons flow and there exists a binding of the electron bands, as seen in Fig. 2-3. In the above case, the electrons of the semiconductor flow to the metal and a space charge builds up at the surfaces. The electrostatic field established by the accumulated space charge is then just enough to prevent any further charge transfer, and the contact potential is given as the difference in the work functions. The electrostatic energy associated with the process is eV_s, where V_s is the total potential drop between the surface and the underlying bulk. Almost the whole contact potential falls across the space charge region, giving

$$-eV_s = W_\phi - W_\phi' = W_\phi - \chi - W_b$$

It follows that V_s depends on the work function of the metal and on the affinity and conductivity of the semiconductor. Germanium and silicon are found not to obey this dependence. The existence of certain surface states has been proposed to explain their behavior.

Krupp (1971) has considered an analysis of metals and semiconductors from a surface state viewpoint. Figure 2-2 should be noted for an analysis of an n-type semiconductor. The *surface* in this case carries a negative surface charge compensated by a positive *space* charge layer. The semiconductor then has a surface double layer. The distance ϕ_0 measures the energy range within which the surface states are filled. Assuming a constant density per unit energy of surface levels $\mathcal{D}_s$, when the semiconductor contacts a metal, charge exchange occurs almost entirely within the surface states. With the gap of contact $Z \approx 4$ Å between the metal and semiconductor, the surface

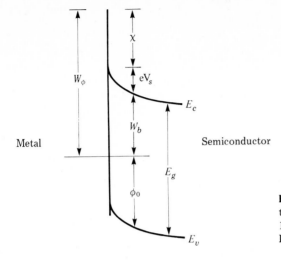

Figure 2-3 Contacted energy bands between metal and semiconductor (Krupp, 1971). Reproduced by permission of the Institute of Physics.

charge density σ on the semiconductor in contact with the metal having a work function W_ϕ is

$$\sigma \cong e \, \mathscr{D}_s \, \frac{W_\phi - E_g - \chi + \phi_0}{1 + e^2 Z \mathscr{D}_s / E_0} \qquad (2\text{-}7)$$

where E_0 = induction constant
e = electronic charge

The work function for the semiconductor is

$$W_{sc} = E_g + \chi - \phi_0 \qquad (2\text{-}8)$$

The surface charge density is thus directly proportional to the difference between the work functions of the two materials. The surface charge density σ from Eq. 2-7 reduces to the capacitor equation (charge = capacitance × voltage) with $e^2 \mathscr{D}_s$ representing the capacitance of the metal-semiconductor interface. Materials such as Si and GaAs are covalent semiconductors having large surface state densities.

Contacts between two semiconductors have been studied by Sze (1969), but less is known about these materials. In all studies one must know the conditions of the surface—whether clean, contaminated, or covered by a monolayer. It should also be noted that insulators can be treated as semiconductors with wide band gaps.

Calculation of electrical adhesion force In order to calculate the electrical adhesion force, the charge distribution at the interface must be determined. This requires the solution of Poisson's equation in one dimension as

$$\frac{\partial^2 V}{\partial x^2} = -\frac{\rho_q}{\kappa} \qquad (2\text{-}9)$$

where ρ_q is the volume charge density and κ is the permittivity. Here, V is the electrostatic potential. The charge density can be given by the distribution

$$\rho_q(t) = \rho_q(0) \exp(-\frac{t}{\tau})$$

(2-10)

with $\tau = \kappa/\sigma_e$, where σ_e represents the electrical conductivity. The solution of this system has the general form $V'' \approx f(V)$. However, a closed-form solution is seldom possible. Auxiliary conditions for solution are charge neutrality of the system and vanishing field strength at a large distance from the interface.

Considering the two surfaces in contact with a potential difference V. The force to transfer a charge between them is given as

$$F = \frac{qV}{2Z_0}$$

(2-11)

where Z_0 is the distance between the surfaces.

From the capacitor equation the charge can be expressed as

$$q = \frac{\kappa_1 VS}{Z_0}$$

(2-12)

Inserting Eq. 2-12 into Eq. 2-11, one obtains the electrical adhesion force per unit area, pressure, as

$$P = \frac{\kappa_1 V^2}{2Z_0^2}$$

(2-13)

In this expression, V is the electrostatic potential obtained from the former analysis.

The surface charge density can also be related to this analysis, which may be beneficial when dealing with certain types of semiconductors. Equation 2-13 expressed in terms of charge density can be written as

$$P = \frac{\sigma^2}{2\kappa}$$

(2-14)

For a metallic spherical adherent and a grounded metallic half-space, Russell (1909) calculated the adhesion force due to electrostatics as

$$F^0_{el} = \frac{q^2}{16\pi\kappa \left[\gamma' + \frac{1}{2} \ln(2R/Z)\right] 2RZ}$$

(2-15)

where q = charge
  γ' = 0.5772 (Euler's constant)

For nonmetallic adherents the sphere is divided up into concentric shells with the charge given as

$$q^2 = q_0^2 \, e^{-x/\sigma'}$$

(2-16)

Table 2-2 Adhesion force and charge of SiC on steel surfaces (Zimon, 1969)

Factor				Value			
D_p, μm	111	95	73	63	56	35	25
F, N × 10^{-5}	3	11	13	10	17	90	120
q/A, C/m^2 × 10^{-7}	1.1	3.9	5.7	6.8	12.1	135	274

Reproduced with permission of Plenum Publishing Corp.

where σ' is the boundary layer thickness with values ranging from 0 to 1 μm. Using this information, the adhesion force for this system is

$$F^0_{el} = \frac{q^2}{16\pi\kappa R\sigma'} \cdot \frac{\ln(1 + \sigma'/Z)}{[\gamma' + \frac{1}{2}\ln(2R/Z)][\gamma' + \frac{1}{2}\ln(2R/Z + \sigma')]} \tag{2-17}$$

Zimon (1969) has experimentally determined the adhesion force of silicon carbon on steel surfaces in nitrogen at varying charge values. Table 2-2 shows these results, indicating that as the particle size decreases the electrical charge increases so the electrical component of the adhesive force becomes greater. The magnitude of the electric charge in adhesion is given in Example 2-2.

Example 2-2 How much charge is necessary to maintain an iron particle of size 10 μm on a polished copper surface with a 50 × 10^{-8} N force?

$$F^0_{el} = 50 \times 10^{-3} \text{ V} \cdot \text{A}$$

$$= 50 \times 10^{-8} \text{ V/m}$$

$$F^0_{el} = \frac{q^2}{16\pi(8.86 \times 10^{-12} \text{ A} \cdot \text{sec/V} \cdot \text{m})} \cdot$$

$$\frac{1}{(0.5772 + \frac{1}{2}\ln[(10 \times 10^{-6})/(4 \times 10^{-10})]^2)(40 \times 10^{-16})}$$

$$50 \times 10^{-8} = \frac{q^2}{5.66 \times 10^{-23}}$$

$$q = 5.32 \times 10^{-15} \text{ C}$$

Capillary Forces

As the humidity of the air increases, the importance of electrostatics is generally seen to wane, while the adhesion due to the adsorbed liquid films between the solid particles and adhering surface increases. This condensed water film forms a gap or bridge between the particle and the surface.

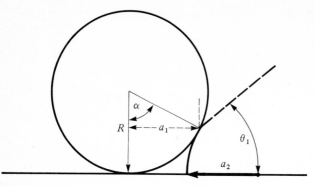

Figure 2-4 Contact angle representation between two solids and a liquid gap.

In order to analyze the capillary forces existing on a solid particle in contact with a surface with an enclosed liquid gap as seen in Fig. 2-4, one can look at the free energy of such a system ($\mathscr{F}$). This free energy at equilibrium has a minimum value. The surface free energy at equilibrium can be written as

$$d\mathscr{F} = 0 = -p_\gamma \, dV + dS \tag{2-18}$$

where p_γ = capillary pressure
  S = surface area
  V = volume
  γ = interfacial tension

The capillary pressure can be related to the geometry of the system and the interfacial tension. Laplace has performed these calculations, yielding

$$p_\gamma = -\gamma\left(\frac{1}{a_2} - \frac{1}{a_1}\right) \tag{2-19}$$

where a_1 and a_2 are the distances defined in Fig. 2-4.

Considering the force balance between the meniscus of liquid and the particle and the surface, the force of the interfacial tension draws on the particle and is opposed by the liquid pressure due to the concave shape, producing

$$F_{cap} = F_{IFT} - F_{LP} \tag{2-20}$$

The interfacial tension force is given as

$$F_{IFT} = 2\pi a_1 \gamma \tag{2-21}$$

and the liquid pressure force as

$$F_{LP} = S p_\gamma \tag{2-22}$$

with $S = \pi a_1^2$, the area of contact.

Inserting these expressions into Eq. 2-20 and with the help of Eq. 2-19, one has

$$F_{cap} = 2\pi a_1 \gamma + \pi a_1^2 \gamma \frac{a_1 - a_2}{a_1 a_2} \qquad (2\text{-}23)$$

By geometric consideration from Fig. 2-4,

$$a_1 = R(2 + \tan \alpha - \sec \alpha)$$

$$a_2 = R(\sec \alpha - 1)$$

These geometric factors reduce the capillary force equation to

$$F_{cap} = \frac{2\pi \gamma R}{1 + \tan \alpha/2} \qquad (2\text{-}24)$$

If one of the surfaces is planar, the height of the intermediate layer is twice as small, giving the capillary force as

$$F_{cap} = 4\pi \gamma R \qquad (2\text{-}25)$$

This assumes total wetting of the smooth surfaces. If θ_1 is the wetting angle, the force is modified by this factor:

$$F_{cap} = 4\pi \gamma R \cos \theta_1 \qquad (2\text{-}26)$$

In practice, the equations for predicting the adhesion due to capillary forces are approached only at relative humidities close to 100%. The wetting angles for various systems are shown in Example 2-3.

Example 2-3 The wetting angle between an alumina particle on a smooth glass surface varies with liquid wetting the surfaces. For a 100-μm-diameter alumina particle, what is the wetting angle for the measured adhesion forces with various liquids listed in Table 2-3? Assume air saturation conditions with each liquid.

Table 2-3

F_{cap}, μN	Liquid	γ, mN/m	θ, calculated
20	Water	72.7	64°
15	Glycerin	63.5	68°
5	Decane	25	71°
4	Octane	21.8	73°

Using Eq. 2-2,

$$\cos \theta = \frac{F_{cap}}{4\pi \gamma R}$$

$$\theta = \cos^{-1}\left(\frac{F_{cap}}{4\pi \gamma R}\right)$$

Particle Shape and Surface Conditions

The area of contact is a crucial but elusive parameter in calculating adhesion, especially in the electrostatic component. No direct measurement exists for the determination of this area. The value changes, depending on the adhering particle and the substrate onto which it adheres. Morgan (1961) has suggested a formula for determination of the radius of the contact area. This formula is dependent on the applied force, Poisson's ratios, and elastic moduli of the materials:

$$R_c = \sqrt[3]{0.75 \, RF \left[(1 - \mu_1^2)/E_1 + (1 - \mu_2^2)/E_2 \right]} \tag{2-27}$$

where R = radius of the particle
              F = applied force
        μ_1, μ_2 = Poisson ratios
          E_1, E_2 = elastic moduli

Example 2-4 applies the above expression to determine areas of contact.

Example 2-4 What percentage of the area of a coupon 10^{-4} m^2 is covered by the contacting of aluminum particles having a 2×10^{-5} N force applied to them on a copper surface? The size distribution is given in Table 2-4.

$$\mu_{Al} = 0.36$$

$$\mu_{Cu} = 0.33$$

$$E_{Al} = 6.89 \times 10^{10} \text{ N/m}^2$$

$$E_{Cu} = 10.3 \times 10^{10} \text{ N/m}^2$$

Equation 2-27 can be used:

$$R_c = \left[0.75RF \left(\frac{1 - \mu_1^2}{E_1} + \frac{1 - \mu_2^2}{E_2} \right) \right]^{\frac{1}{3}}$$

$$= \left[0.75R(2 \times 10^{-5} \text{N}) \left(\frac{1 - 0.36^2}{6.89 \times 10^{10}} + \frac{1 - 0.33^2}{10.3 \times 10^{10}} \right) \right]^{\frac{1}{3}}$$

Table 2-4

D_p, μm	Number of particles
200	5
150	12
100	10
50	30
20	20
10	5

Substituting the values for R and using the formula area $= (\pi/4)(2R_c)^2$, one finds:

R, μm	R_c, m	Area of one particle, m^2	Number of particles	Area$_T$, m^2
100	0.317×10^{-6}	0.316×10^{-12}	5	1.58×10^{-12}
75	0.288×10^{-6}	0.261×10^{-12}	12	3.13×10^{-12}
50	0.252×10^{-6}	0.199×10^{-12}	10	1.99×10^{-12}
25	0.199×10^{-6}	0.124×10^{-12}	30	3.72×10^{-12}
10	0.1471×10^{-6}	0.068×10^{-12}	20	1.36×10^{-12}
5	0.116×10^{-6}	0.042×10^{-12}	5	0.21×10^{-12}
				12.0×10^{-12}

A small fraction of the total surface area is in actual contact with the particles.

As often mentioned before, the shape of the particle is of importance in adhesion as well as other phenomena of particle behavior. Flat platelets would have more surface contact than spherical or jagged particles. The sphericity given previously is one way of classifying this shape term. Another method is the procedure of Heywood (1963). Figure 2-5 shows the adhesion force F_{ad} for a particular system with varying sphericity factors.

Generally, minimum adhesion forces occur for particles of isometric shape, that is, particles that look more like spheres or regular polygons. Planar particles have higher adhesion forces than isometric shapes. One also finds the adhesion forces of needles and fibers to be larger than both isometric and planar particles.

Turning from the particle and its shape and area of contact, many other surface conditions on the substrate are possible. The ratio of the size of the adhering particles and the surface roughness measured as the height of hills and depth of valleys on the surface shows some unique behavior.

A designer can control the surface finish on a metal by specifying the desired condition. As time passes, the surface condition may change due to oxidation, corrosion,

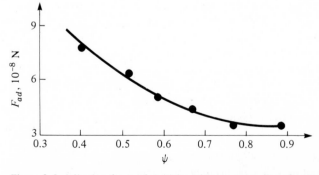

Figure 2-5 Adhesion force of particles with a mean radius of approximately 130 μm as a function of the sphericity factor (Zimon, 1969). Reproduced with permission of Plenum Publishing Corp.

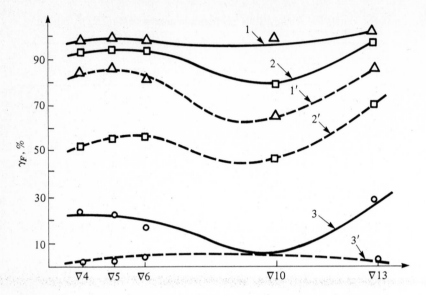

Figure 2-6 Adhesion numbers (percentage of particles remaining on the surface) of spherical glass particles of various diameters on steel surfaces of various types of surface finishes with detaching forces of 0.07 kg (1, 2, 3) and 1.15 kg (1', 2', 3') (Zimon, 1969). Reproduced with permission of Plenum Publishing Corp.

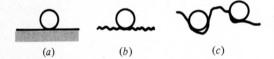

(a)   (b)   (c)

Figure 2-7 Types of substrate roughness associated with the adhesion of particles (Zimon, 1969). Reproduced with permission of Plenum Publishing Corp.

and erosion, so that what adheres to a surface in its new state may be completely different from what adheres to the surface as time goes on. Zimon (1969) showed the adhesion behavior of glass particles on steel surfaces of varying finishes. There is an obvious minimum in adhesion with surface roughness, as seen in Fig. 2-6. Three conditions can be designated as occurring to produce this minimum behavior. Figure 2-7 shows these states. State (a) is when the surface is ideally smooth, produced by a series of polishing steps. State (b) is when microscopic indentations or asperities on the surface are of a small order of magnitude in comparison to the particle size. This condition has smaller contact area than the smooth condition, state (a), and the state (c). State (c) is seen to occur when the size of the roughness is macroscopic so that the particles fall into the indentations, thus sizably increasing the contact area. Table 2-5 shows the surface roughness of several common materials and finishes. Various processing operations are related to adhesion forces in Example 2-5.

Example 2-5 Comment on the qualitative size of the adhesion force in relation to the surface roughness and particle sizes listed below Table 2-5.

Table 2-5 Surface roughness by common production methods

Process	Roughness height, μm
Sawing	25.4–0.813
Drilling	6.35–1.6
Milling	6.35–0.813
Reaming	3.18–0.813
Grinding	1.6–0.010
Polishing	0.406–0.010
Superfinishing	0.0203–0.010

Metalworking, January (1964). Reproduced by permission of Mac-can Publishing Co.

Surface finish	Particle diameter, μm
Sawing, 25.4 to 0.81 μm	500
Milling, 6.35 to 0.81 μm	150
Polishing, 0.41 to 0.1 μm	100, 10
Superfinishing, 0.21 to 0.1 μm	25, 2

	Diameter/finish height
Sawing	19.6 to 617
Milling	23.6 to 185
Polishing	243 to 1000, 24.3 to 100
Superfinishing	125 to 250, 10 to 20

In these cases the diameter is much greater than the finish height, showing little of the effect seen in Fig. 2-7c.

Temperature

Little work has been done on the effect of temperature on adhesion. It has been found in some studies that as the temperature increases, the adhesion also increases. This may be attributed to a number of factors. Higher temperatures tend to drive off the adsorbed water layers and the electrostatic forces may be more dominant. In decreasing the temperature below 0°C one finds little change in adhesion, since the water adsorbed probably freezes the particle to the surface, showing no change in adhesion as the applied forces for dislodging the particles are increased.

For painted surfaces, increased temperatures make the surface sticky and thus increase the adhesional forces.

Contact Time

The length of time particles remain on a surface seems to have little effect after a maximum of 30 min, according to some experimental evidence (Deryagin and Zimon,

1961). If the conditions under which the particles and substrate are stored were to change, it is difficult to ascertain what would happen. Ideally, the samples should be stored under constant humidity and temperature in a clean gas container. Often, this is a difficult condition to maintain. In an actual application of adhesion forces, the conditions in an industry or process probably vary from day to day or continually during the process.

2-3 EXPERIMENTAL TECHNIQUES

There are several techniques available for the measurement of the adhesion force of particles on substrates. These techniques involve both the study of individual particles (as in the microbalance technique) and of multiple particles by statistical counting (as in the centrifuge method). In each type of study the conditions of both the particles and the substrates must be known with high precision. The exact chemistry, physical size and shape, surface condition, temperature, and properties of the surrounding gas are vital parameters that must be quantified in order to obtain reproducibility.

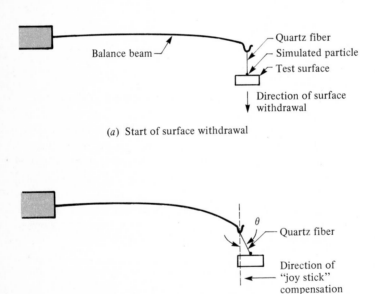

(a) Start of surface withdrawal

(b) End of surface withdrawal prior to break

Figure 2-8 Microbalance technique using quartz fibers (Corn, 1961). Reproduced by permission of *J. Air Pollution Control.*

The microbalance technique is shown in Fig. 2-8. This method is based on the use of quartz fibers, which can form a cantilever spring or torsion balance. Sensitivities to 0.1 μg are possible with an electromicrobalance. Several investigators have used this apparatus (for example, Corn, 1961; Hamaker, 1957; Overbeek and Sparnaay, 1954; Stone, 1930). As with any sensitive technique, extreme care must be taken for accurate results.

The pendulum method also employs the use of one particle. The particle is suspended on a vertical fiber in contact with the substrate. The substrate is then raised and the angle between the fiber and the vertical plane is recorded at the moment of breakage of the particle-surface adhesion force. This technique is generally applicable only for relatively large particles.

The centrifuge method is probably the most used technique to study the adhesion of particles to substrates. Generally, high speed centrifuges (up to 25,000 rpm) and ultracentrifuges (with rotations up to 100,000 rpm or more) have been used for adhesion studies. A special container to hold the substrate sample has generally been constructed. The temperature and humidity of the container can be controlled if the system is sealed tightly. Microscopic analysis in place would be desirable, but this has not yet been achieved. The force required to remove the particles in a centrifugal field is

$$F = \frac{\pi D_p^3}{6} \rho_p \omega^2 R_{CL} \qquad (2\text{-}28)$$

where ω = angular speed, rad/sec
  R_{CL} = distance from the center of the centrifuge axis
  ρ_p = density of particle

Techniques are now possible to analyze a series of fields on the substrate having a few thousand particles.

The aerodynamic technique uses a flow field or air to dislodge a particle adhering to the surface. Generally, high-velocity air jets have been employed. This technique suffers from the difficulty in measuring the exact aerodynamic conditions in the vicinity of the adhering particles.

The inclined-plane method of finding the adhesion force measures the friction between a particle bed and a substrate by varying the angle of inclination until the particles begin to slide. Varying powered bed thicknesses have been studied. By plotting the normal pressure against the frictional force and extrapolating to zero normal force, the adhesion between particles and substrate is obtained. This technique relies on the relationship between adhesion and friction and the complicated nature of the sliding process, which is not well known.

Table 2-6 contains some experimental values for the adhesion forces of different particles to various substrates in order to give the reader an order of magnitude appreciation of the nature and size of the adhesional force. The results of a study on the adhesion of coal to carbon steel are shown in Example 2-6.

Table 2-6 Adhesion forces of selected materials at 50% relative humidity (Kordecki and Orr, 1960)

Particle	Solid surface	Force, mN
Glass	Aluminum	1.35
Glass	Copper	4.18
Glass	Plexiglas	1.44
Sand	Glass	0.76
Charcoals	Glass	0.55
Glass	Glass	0.37

Reproduced by permission of Archives of Environ. Health.

Example 2-6 The adhesion of coal to a polished carbon steel surface was studied in a Sorvall centrifuge. At a rotational speed of 15,000 rpm, the size distribution given in Table 2-7 was obtained by a microscopic analysis. The coal is a Montana lignite having a specific gravity of 1.20. The distance from the center of the centrifuge to the sample is 4.2×10^{-2} m. With this information, calculate the adhesion force based on the length average particle size of the particles remaining on the carbon steel surface.

Using Table 1-2, one can find $D_{pl} = 11.81$ μm.

$$\text{Force} = \left[\rho_p \left(\frac{\pi}{6} \right) D_{pl}^3 \right] \text{kg} \left[\left(\frac{2\pi\, 15{,}000}{60} \right)^2 \, 4.2 \times 10^{-2} \right] \text{m/sec}^2$$

$$= 1.63 \times 10^9 D_{pl}^3 \text{ N}$$

$$= 2.68 \times 10^{-6} \text{ N}$$

Table 2-7

D_p, μm	Number of particles remaining on surface
1	922
2	888
3	776
4	654
5	526
6	415
7	320
8	274
9	240
10	212
15	75
20	27
25	9
30	3

2-4 AGGLOMERATION

Introduction

The term *agglomeration* has been used rather freely in the literature to mean the combination—whether weak or strong—of two or more particles to form one (stable or unstable) particle. Opoczky (1977) has attempted to classify agglomeration more systematically. Opoczky suggests that weak reversible adhesion of particles due to van der Waals forces should be called *aggregates* and the overall process should be called aggregation. Forces range from 0.04 to 4 kJ/mol in the aggregation area. Agglomeration is an irreversible combination of particles by chemical forces varying from 40 to 400 kJ/mol. Since this classification is relatively new, the literature is filled with essentially what authors of individual works define agglomeration to be.

Agglomeration is a very important aspect of handling in gas-solid flow. In transport lines and fluidized beds, one wants to get rid of it, while in cyclones and electrostatic separators, one takes advantage of its occurrence so that equipment works properly. In turbulent flow systems with a polydispersity of particles, the probability of agglomeration is high due to the various trajectories that exist for polydispersed systems. Monodispersed systems have lower probabilities of agglomeration but not significantly lower. Sometimes the flow models do not behave according to the static particle size distribution obtained in analyzing the flow system. The models seem to work best when large particles are assumed. Often, agglomeration is assumed and the system is treated as though it has large particles. Capes (1974) analyzed high-pressure gas fluidization of fine particles and modified the Richardson–Zaki model for the superficial gas velocity related to bed porosity by modifying the exponent term on the porosity and suggesting that this change is due to agglomeration. In reality, one can never be sure this agglomeration is happening unless visual experiments can be performed to ascertain the presence of agglomeration. Little information exists on this topic to date.

Arundel, Bibb, and Boothroyd (1970) found agglomeration of solids to increase near the wall in their study of a 1- to 40-μm-diameter zinc particles flowing in a vertical tube. They also found that the turbulence was suppressed near the wall, contributing to the agglomeration phenomenon.

Agglomeration of particles in a flowing system was also investigated by Fink (1975), who observed transmission levels of a light signal. As the solids concentration went from 0.05 to 0.5 kg/m^3, the particles of 8.6 μm diameter increased in size from 20 to 200 μm diameter.

Mechanisms

Four different mechanisms for agglomeration have been suggested by Sastry and Fuerstenau (1973):

1. Coalescence
2. Breakage

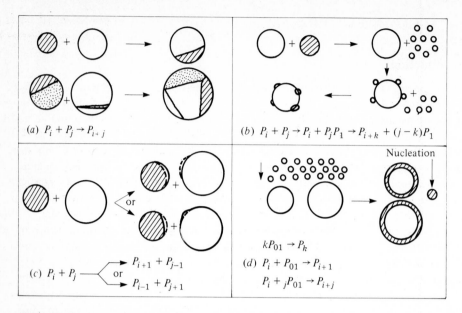

Figure 2-9 Representation of growth mechanisms: (*a*) coalescence, (*b*) breakage, (*c*) abrasion transfer, (*d*) snowballing and nucleation. The symbol P_1 represents a green pellet of mass m_f (= *ih*); P_{01} represents a freshly added nuclei of mass *h* (Sastry and Fuerstenau, 1973). Reproduced by permission of Powder Technology, Elsevier Sequoia S.A.

3. Abrasion transfer
4. Snowballing and nucleation

Figure 2-9 illustrates these mechanisms. In coalescence, large particles are produced by the clumping of two or more colliding granules; only binary coalescence is considered here for simplicity. The breakage sequence leads to daughter fragments of the same or different sizes. In abrasion transfer a certain mass of material is transferred from one specie to another. Snowballing or layering is the combination of several nuclei of particles to form a larger particle. These mechanisms can occur in a flow system as well as in the grinding process for which they were initially suggested.

Binding Forces

There are several binding mechanisms to form agglomerates; some of these are reversible and some are more or less irreversible. The mechanisms that are considered reversible are the same mechanisms that come into play in the adhesion area—van der Waals and electrostatic forces. Instead of viewing adhesion between particulate matter and a planar surface, one now views the interparticle mechanism. The van der Waals and electrostatic forces can be calculated for this interparticle agglomeration. In the van der Walls force case,

$$F_{VDW} = \frac{\pi}{2}\left(\frac{3 \times 10^{-19}}{Z^2}\right)\left(\frac{D_{p1}D_{p2}}{D_{p1} + D_{p2}}\right) \qquad (2\text{-}29)$$

where F_{VDW} is in newtons and Z is the distance in meters (Orr, 1966).

For large particles these reversible forces are indeed small, but for small particles these forces may become dominant.

Solid bridging This involves the direct fusion between particles at contact points. Fusion may be caused by melting, by solid diffusion, through binding agents, by chemical reactions, or by crystallization. Large solid bridges are formed in plastic solids. Increased temperature and pressure may cause accelerated bridging.

Liquid bridging This binding action may be considered a liquid bridge force. When acting on several particles collectively, this force can form sizable agglomerates, and gas can be trapped in such agglomerates. Rumpf (1962) calculated the tensile strength of a capillary-state agglomerate as

$$T_c = \left[\frac{6}{D_p}\left(\frac{1-\epsilon}{\epsilon}\right) + \frac{C_c}{A}\right]\gamma \cos\theta_1 \qquad (2\text{-}30)$$

where C_c = agglomerate circumference
A = cross-sectional area of agglomerate
γ = interfacial tension
θ_1 = wetting angle
ϵ = volume fraction of voids

In order to increase the binding action of a liquid its viscosity can be increased. Some high-viscosity liquids used as binders are water solutions with sugar, glue, gelatin, dextrons, gums, starch, and molasses. Example 2-7 shows the magnitude of the tensile strength of an agglomerate.

Example 2-7 Consider an agglomerate held together by a liquid water bridge. The size of the agglomerate is 500 μm diameter and it is composed of 25-μm-diameter particles. The void fraction has been determined to be 0.30, with a 70° wetting angle. Find the tensile strength of the agglomerate. Using Eq. 2-30, one finds:

$$T_c = \left[\frac{6}{25 \times 10^{-6}\text{ m}}\left(\frac{0.7}{0.3}\right) + \frac{\pi 500 \times 10^{-6}\text{ m}}{(\pi/4)(500 \times 10^{-6})^2\text{ m}^2}\right](72 \times 10^{-3}\text{N/m})\,0.34$$

$$= 1.1 \times 10^8\text{N/m}^2$$

PROBLEMS

2-1 For a ceramic 100-μm-diameter particle on a carbon steel surface, estimate the size of the adhesion force using Lifshitz's and Hamaker's theories.

2-2 In a recent study on the adhesion of coal particles to common metal surfaces at 70% relative humidity, the adhesion force was found to be largest for carbon steel, followed by aluminum, copper, and stainless steel. Can you explain this ordering of adhesion forces?

2-3 Calculate the size of the contributing forces to the adhesion of a 10-μm-diameter glass sphere to an aluminum oxide surface, considering van der Waals, electrostatic, and capillary forces are present. Assume the charge on the system has been approximated at 10^{-14} C.

2-4 Considering the capillary force only on spherical glass particles on a carbon steel surface at an air relative humidity of 65%, the following experimental data were compiled:

Factor		Value		
D_p, μm	50	25	15	7
F_{ad}, N	2.1×10^{-9}	2.1×10^{-8}	6.1×10^{-8}	1.3×10^{-7}

Using a $0°$ contact angle, determine the adhesion force as given by the theoretical development and compare this to the experimental findings.

2-5 In an experiment investigating alumina adhering to a glass surface over a range of particle sizes from 50 to 10 μm no difference was seen in adhesion force as the particle size changed. Interpret these findings using the basic equations for adhesion theories.

2-6 In the study of coal adhesion to an aluminum surface the data in the table were obtained by use of the centrifuge technique. Analyze these findings and make appropriate conclusions.

	Number of particles			
	Field 1	Field 2	Field 3	Field 4
At 5000 rpm				
1	609	589	740	777
2	594	555	709	779
3	573	537	655	680
4	486	485	600	573
5	417	436	478	499
6	348	396	429	454
7	288	325	350	370
8	259	293	295	325
9	238	246	253	273
10	185	232	217	220
11	169	224	189	191
12	129	185	150	155
13	113	163	128	127
14	102	130	111	118
15	81	119	99	101
16	72	110	86	96
17	70	93	66	77
18	63	79	48	74
19	50	75	41	61
20	46	69	35	49
25	29	39	13	26
30	17	15	5	13
35	12	5	3	3
40	7	3	1	1
45	2	1	0	0

	Number of particles			
	Field 1	Field 2	Field 3	Field 4
At 15,000 rpm				
1	892	521	454	631
2	844	507	450	608
3	811	503	431	588
4	727	480	418	556
5	597	446	395	540
6	485	409	354	465
7	425	366	319	407
8	362	318	300	365
9	330	274	266	347
10	281	343	249	293
11	260	229	222	272
12	211	198	194	236
13	170	168	175	205
14	164	146	150	164
15	129	127	142	154
16	112	118	122	130
17	95	99	107	111
18	85	84	96	105
19	67	70	81	89
20	61	64	74	77
25	26	31	48	51
30	15	19	31	31
35	4	9	20	16
40	1	5	14	8
45	0	3	9	3
50	0	2	5	2
75	0	0	1	0

REFERENCES

Arundel, P. A., S. D. Bibb and R. G. Boothroyd: *Powder Technol.* **4**:302 (1970).

Capes, C. E.: *Powder Technol.* **10**:303 (1974).

Corn, M.: *J. Air Pollution Contr. Assoc.* **11**:566 (1961).

——: Chapter XI in C. N. Davies (ed.), *Adhesion of Particles in Aerosol Science,* Academic Press, London, 1966.

Deryagin, B. V., and A. D. Zimon: *Kolloid Zh.* **23**:454 (1961).

Fink, Z. J.: *Powder Technol.* **12**:287 (1975).

Hamaker, H. C.: *Physica* **4**:1058 (1937).

Heywood, H.: *J. Pharm. Pharmacol. Suppl.* **15**:56T (1963).

Kittel, C.: *Introduction to Solid State Physics,* Wiley, New York, 1971.

Kordecki, M. C., and C. Orr, Jr.: *Arch. Environ. Health* **1**:1311 (1960).

Krupp, H.: *Adv. Colloid Interface Sci.* **1**:111 (1956).

——: *Static Electrification,* Institute of Physics, London, 1971.

Loeb, L. B.: *Static Electrification,* Springer, Berlin, 1958.

Metalworking (Jan. 1964).

Morgan, B. B.: *Brit. Coal Utilization, Res. Assoc. Monthly Bull.* **25**:125 (1961).

Opoczky, L.: *Powder Technol.* **17**:1 (1977).

Orr, C., Jr.: *Particulate Technology,* Macmillan, New York, 1966.

Overbeek, J. T. G., and M. Sparnaay: *Discuss. Faraday Soc.* **18**:12 (1954).

Rumpf, H.: In W. A. Knepper (ed.), *Agglomeration,* Interscience (Wiley), New York, 1962.

Russell, A.: *Proc. R. Soc.* **A82**:524 (1909).

Sastry, K. V. S., and D. W. Fuerstenau: *Powder Technol.* **7**:97 (1973).

Stone, W.: *Phil. Mag.* **9**:610 (1930).

Sze, S. M.: *Physics of Semiconductor Devices,* Wiley, New York, 1969.

Zimon, A. D.: *Adhesion of Dust and Powders,* Plenum, New York, 1969.

THREE

PARTICLE DYNAMICS AND TURBULENCE

3-1 INTRODUCTION

The behavior of single particles in fluids generates many interesting physical phenomena and paradoxes. In his books and films on shape and flow of particles, Shapiro (1961) has commented on both these aspects. One would intuitively expect that as the velocity of a particle increases, the drag should increase. In general, this is true, but the transitional flow region presents some exceptions to this trend. For particles, one finds that for low-velocity flow the pressure forces dominate, while for higher turbulent flows the viscous forces take over. The surface condition of the particles may profoundly influence the particle drag in a flow situation. A roughened particle produces a turbulent boundary layer at smaller speeds than a smooth particle. The turbulent boundary layer has a smaller wake behind it and thus less drag. This drag may be even less than that of laminar flow over a smooth particle.

3-2 BASIC PARTICLE DYNAMIC MODEL

The general equation of motion is based on treatments of particle dynamics by Basset, Boussinesq, and Oseen and is essentially an application of Newton's second law (Soo, 1967):

$$\overbrace{\left(\frac{\pi D_p^3 \rho_p}{6}\right)\frac{dU_p}{dt}}^{A} = \overbrace{\left(\frac{\pi D_p^3 \rho_p}{6}\right)\left(\frac{3C_D \rho_f}{4 D_p \rho_p}\right) |U_f - U_p|(U_f - U_p) - \left(\frac{\pi D_p^3}{6}\right)\frac{\partial p}{\partial x}}^{B} \overbrace{}^{C}$$

$$+ \overbrace{\frac{1}{2}\left(\frac{\pi D_p^3}{6}\right)\rho_f \frac{d}{dt}(U_f - U_p)}^{D}$$

$$+ \overbrace{\left(\frac{3D_p^2}{2}\right)(\pi \rho_f \mu_f)^{1/2} \int_{t_0}^{t} \frac{(dU_f/dt') - (dU_p/dt')}{\sqrt{t - t'}} dt'}^{E} + \overbrace{F_e}^{F} \qquad (3\text{-}1)$$

The lettered terms can each be given a physical interpretation:

A = mass × acceleration of the particle, which is present only in unsteady flow situations representing the force necessary to accelerate the particle

B = drag force containing a drag coefficient that is a function of the Reynolds number

C = force from the pressure gradient in the fluid surrounding the particle

D = force due to acceleration of the apparent mass of the particle relative to the fluid

E = Basset force, which is the force due to the deviation of the flow pattern around the particle from steady-state conditions; this depends on the previous motion of the particle and the fluid

F = external force

Tchen (1947) has represented term C as

$$\left(\frac{\pi D_p^3}{6}\right)\rho_f \left(\frac{dU_f}{dt}\right) \qquad (3\text{-}2)$$

while Corrsin and Lumley (1956) suggest that C be given as

$$\left(\frac{\pi D_p^3}{6}\right)\left[\rho_f \frac{dU_f}{dt} - \mu_f \left(\frac{\partial^2 U_f}{\partial x^2} + \frac{\partial^2 U_f}{\partial y^2} + \frac{\partial^2 U_f}{\partial z^2}\right)\right] \qquad (3\text{-}3)$$

using the Navier–Stokes equation in rectangular cartesian coordinates. Boothroyd (1971) has made several observations on the order of magnitude of certain terms in Eq. 3-1 for gas-solid systems. For large values of ρ_p/ρ_f, which is common in gas-solid systems except possibly for very high pressure flow systems, the terms C, D, and E are small compared to A and B. Of course, at steady state term A is zero, leaving only B and F of importance. Torobin and Gauvin (1961) have analyzed the unsteady particle flow extensively. Odar (1966, 1968) has attempted to modify Eq. 3-1 for turbulent flow by the use of empirical coefficients.

For the case of turbulent flow for particle Reynolds number $\ll 1$ Chao (1964) and Hjelmfelt and Mockros (1966) have considered Eq. 3-1 as

$$\frac{dU_p}{dt} = a_1(U_f - U_p) - b_1 \frac{dU_f}{dt} - c_1 \int_{t_0}^{t} \frac{(dU_f/dt') - (dU_p/dt')}{\sqrt{t - t'}} \, dt' \qquad (3\text{-}4)$$

where $a_1 = \dfrac{18\mu_f}{(\rho_p + \rho_f/2)D_p^2}$

$$b_1 = \frac{3}{1 + 2\rho_p/\rho_f}$$

$$c_1 = \frac{9(\mu_f/\rho_f\pi)^{1/2}}{1/2 + \rho_p/\rho_f}$$

Various investigators have solved this equation assuming different terms to be zero. The magnitudes of the acceleration times of particles using Eq. 3-1 are shown for specific cases in Examples 3-1 and 3-2.

Example 3-1 Consider the motion of a 1000-μm-diameter coal particle with specific gravity $= 1.2$ being accelerated with air from an initial velocity of 0.305 m/sec to a steady state in an airstream having a temperature of 21°C and 2027 kN/m^2 pressure. Assume at all times that $U_p = \frac{3}{4} U_f$. Find the time necessary to accelerate this particle to a condition where $U_f = 15.24$ m/sec. Use Newton's range drag coefficients over the acceleration period and assume that terms C and D in Eq. 3-1 are significant. Ignore the Basset contribution.

Equation 3-1 simplifies to

$$\left(\frac{\pi D_p^3 \rho_p}{6}\right) \frac{dU_p}{dt} = \left(\frac{\pi D_p^3 \rho_p}{6}\right)\left(\frac{3}{4} \frac{C_D \rho_f}{D_p \rho_p}\right)(U_f - U_p)^2 - \left(\frac{\pi D_p^3 \rho_f}{6}\right)\frac{dU_f}{dt}$$

$$+ \frac{1}{2}\left(\frac{\pi D_p^3}{6}\right)\rho_f \frac{d}{dt}(U_f - U_p)$$

Now,

$$\begin{aligned}
D_p &= 1000 \ \mu\text{m} \\
\rho_p &= 1200 \ \text{kg/m}^3 \\
C_D &= 0.44 \ (\text{assuming Newton's range}) \\
\mu_f &= 0.01 \ \text{cP} \\
\rho_f &= 24 \ \text{kg/m}^3
\end{aligned}$$

Then, with $U_p = \frac{3}{4} U_f$,

$$a \frac{dU_p}{dt} = bU_p^2 - c \frac{dU_p}{dt} + d \frac{dU_p}{dt}$$

where $\dfrac{c}{a} = 2.66 \times 10^{-2}$

$\dfrac{d}{a} = 3.32 \times 10^{-3}$

From which one has

$$\frac{dU_p}{dt} = 0.716\, U_p^2$$

$$\int_{0.305}^{11.4 \text{ m/sec}} \frac{dU_p}{U_p^2} = 0.716 \int_0^{t_{final}} dt = -\frac{1}{U_p} \bigg|_{0.305}^{11.4} = 0.716\, t_{final}$$

$$t_{final} = 4.46 \text{ sec}$$

Note that including terms C and D in Eq. 3-1 has minimal effect in this situation.

Example 3-2 A particle in a fluid stream is undergoing an acceleration. Assume that the drag force is the only impedance force. It is desired to calculate the time it takes for particles of various sizes to respond to the acceleration and achieve 63.2% of their final velocity. This time is called the *relaxation time* of the particle. Assume Stokes' law holds for the drag coefficients.

The basic equation can be written as

$$m_p \frac{dU_p}{dt} = C_D \frac{(U_f - U_p)^2}{2} \rho_f A_p$$

where A_p = projected area = $\dfrac{\pi}{4} D_p^2$

m_p = mass of the particle

For Stokes' regime,

$$C_D = \frac{24 \mu_f}{D_p (U_f - U_p)\rho_f}$$

Combining the above,

$$\left(\frac{\pi D_p^3 \rho_p}{6}\right) \frac{dU_p}{dt} = \frac{24 \mu_f}{D_p(U_f - U_p)\rho_f} \frac{(U_f - U_p)^2}{2} \rho_f \left(\frac{\pi}{4} D_p^2\right)$$

and simplifying,

$$\int_{U_{pi}}^{U_{pf}} \frac{dU_p}{U_f - U_p} = \frac{18 \mu_f}{D_p^2 \rho_p} \int_0^t dt$$

$$\frac{(U_f - U_p)_f}{(U_f - U_p)_i} = \exp\left(\frac{-18 \mu_f t}{D_p^2 \rho_p}\right)$$

Table 3-1

D_p, μm	τ', sec $\times$ 0.055
1	10^{-4}
10	10^{-2}
100	1
1,000	10^2
10,000	10^4

When

$$t = \frac{D_p^2 \rho_p}{18 \mu_f}$$

the final particle velocity is 63.2% of the initial particle velocity, and relaxation time is given as

$$\tau' = \frac{D_p^2 \rho_p}{18 \mu_f}$$

For a solid material with specific gravity = 1 and air at standard conditions, the relaxation time varies as given in Table 3-1.

3-3 DRAG COEFFICIENTS

Equation 3-1 contains a drag coefficient C_D, which is a function of the Reynolds number. This drag coefficient is highly dependent on the fluid regimes of laminar and turbulent. For the laminar regime, Stokes (1851) solved the fluid dynamics equations for flow past a sphere, determining the drag force in the sphere as

$$F_D = 3 \, \rho_f \mu_f \, |U_f - U_p| D_p \tag{3-5}$$

with a drag coefficient of

$$C_D = \frac{24}{D_p |U_f - U_p| \rho_f / \mu_f} \tag{3-6}$$

These forces and drag coefficients have been verified experimentally, mainly from measurements on the terminal velocity of a settling particle. The main factors considered in such studies are terms A, B, and F in Eq. 3-1. For the terminal velocity U_t determination, $A = 0$ and $F_e = (\pi D_p^3/6) \, (\rho_p - \rho_f)g$. For the sedimentation system, the velocity of the fluid is zero. Thus, one has

$$\left(\frac{\pi D_p^3 \rho_p}{6}\right) \left(\frac{3}{4} \frac{C_D}{D_p} \frac{\rho_f}{\rho_p}\right) U_p^2 = \frac{\pi D_p^3}{6} (\rho_p - \rho_f)g \tag{3-7}$$

or

$$U_p = U_t = \sqrt{\frac{4}{3} \frac{(\rho_p - \rho_f)g}{\rho_f} \frac{D_p}{C_D}} \tag{3-8}$$

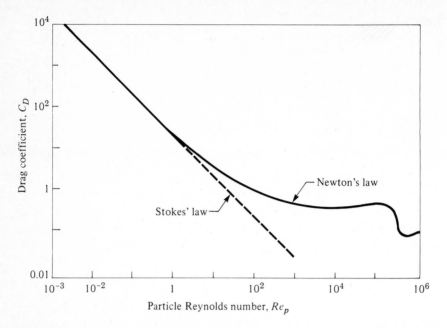

Figure 3-1 Drag coefficient of a sphere as a function of Reynolds number (Boothroyd, 1971). Reproduced by permission of Taylor & Frances Ltd.

The drag coefficient in Eq. 3-8 is given by Eq. 3-6 for Reynolds numbers of the particle less than 2.0. From experimental studies,

$$C_D = \frac{18.5}{Re_p^{0.6}} \quad \text{when} \quad 2 < Re_p < 500 \tag{3-9}$$

and for the range $500 < Re_p < 200{,}000$,

$$C_D = 0.44 \tag{3-10}$$

Figure 3-1 shows the behavior of the drag coefficient as a function of the Reynolds number for spherical particles. The Stokes and Newton ranges are designated. Boothroyd (1971) has compiled a number of expressions for the drag coefficient as a function of the Reynolds number. In each study given here, the shape of the particle was assumed to be spherical. Wadell (1934) considered the drag coefficient for particles of varying sphericity. Figure 3-2 shows this sphericity effect over a limited Reynolds number range, adapted from Wadell's work by Boothroyd.

McCabe and Smith (1976) have suggested a convenient criterion for analysis if the Reynolds number is unknown. Using the terminal velocity value from the equation and the three ranges of drag coefficients, one finds a K factor:

$$K = D_p \left[\frac{g\rho_f(\rho_p - \rho_f)}{\mu_f^2} \right]^{1/3} \tag{3-11}$$

For $K < 3.3$, use Stokes' range; for $3.3 < K < 43.6$ use the intermediate range; and for $43.6 < K < 2360$, use Newton's range.

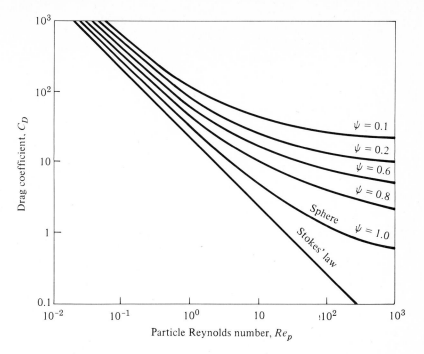

Figure 3-2 Draf coefficients of particles of nonspherical shape (Boothroyd, 1971). Reproduced by permission of Taylor & Frances Ltd.

The terminal velocities of three different particles are presented in Example 3-3. Beyond the value of $K = 2360$ there is an abrupt drop in the drag coefficient due to vortex shedding.

Clift and Gauvin (1970) have closely investigated the motion of particles in turbulent gas flow. The turbulence intensity of the gas stream may drastically affect the drag coefficient of the particle. These investigators define a critical Reynolds number Re_c as the point where the drag coefficient cuts the 0.3 value (see Fig. 3-3). Measurements of drag coefficients in this region have a dependence on the turbulence intensity $I_R = \sqrt{\overline{u'^2}}/(U_f - U_p)$, as shown in Fig. 3-3.

Example 3.3 Determine the settling rates in meters per second for the following situations. Assume the particles achieve their settling velocity quickly.

(a) $D_p = 10,000 \ \mu m$
  $\rho_f = 1.2 \ \text{kg/m}^3$
  $T = 21°C$
  $\rho_p = 1200 \ \text{kg/m}^3$

(b) $D_p = 1000 \ \mu m$
  $\rho_f = 32 \ \text{kg/m}^3$
  $T = 21°C$
  $\rho_p = 1200 \ \text{kg/m}^3$

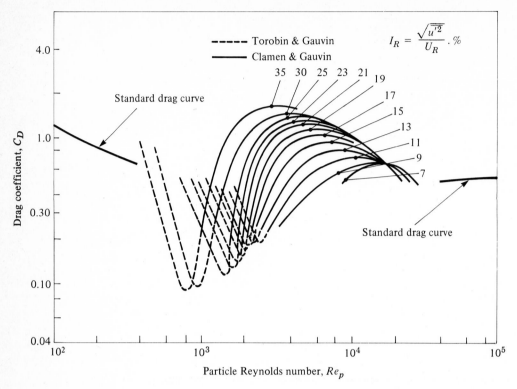

Figure 3-3 Effect of relative intensity of turbulence I_R on the drag coefficients of spheres in the supercritical regime (Clift and Gauvin, 1970). Reproduced by permission of Butterworth Ltd.

(c) $D_p = 100\ \mu m$
$\rho_f = 16\ kg/m^3$
$T = 21°C$
$\rho_p = 1200\ kg/m^3$

Assume the viscosity of the gas in each case is 0.01 cP. First calculate the K values according to Eq. 3-11:

$$K_a = 10^{-2}\left[\frac{9.8 \times 1.2(1200 - 1.2)}{(10^{-5})^2}\right]^{1/3} = 520 \quad \text{(Newton's)}$$

$$K_b = 10^{-3}\left[\frac{9.8 \times 32(1200 - 32)}{(10^{-5})^2}\right]^{1/3} = 154 \quad \text{(Newton's)}$$

$$K_c = 10^{-4}\left[\frac{9.8 \times 16(1200 - 16)}{(10^{-5})^2}\right]^{1/3} = 12.3 \quad \text{(intermediate)}$$

By inserting the drag coefficient from Eqs. 3-9 and 3-10 into the terminal velocity Eq. 3-8, one finds

$$U_{t_{\text{(intermediate)}}} = \frac{0.153 g^{0.71} D_p^{1.14} (\rho_p - \rho_f)^{0.71}}{\rho_f^{0.29} \mu^{0.43}}$$

$$U_{t_{\text{(Newton)}}} = \left[\frac{g D_p (\rho_p - \rho_f)}{\rho_f} \right]^{1/2}$$

Using these equations for (a), (b), and (c), one obtains

$$U_{t_a} = 17.31 \text{ m/sec}$$

$$U_{t_b} = 1.05 \text{ m/sec}$$

$$U_{t_c} = 0.205 \text{ m/sec}$$

3-4 VARIATION ON A SINGLE PARTICLE MOTION

One of the first modifications on the Stokes flow of a single particle was that of Oseen, who accounted for the inertia portion of the fluid dynamics equation by linearizing the term $U(\partial U/\partial x)$ to $U_0(\partial U/\partial x)$. Under this analysis, the drag coefficient is given as

$$C_D = \frac{24}{Re_p} \left(1 + \frac{3}{16} Re_p \right) \tag{3-12}$$

The magnitude of the Oseen correction for a specific case is seen in Example 3-4.

Brenner (1962, 1964, 1966) has given a generalized analysis of Stokes' flow for an arbitrary particle in an infinite medium. In addition, Brenner (1961) has suggested a relation between Oseen drag and Stokes' drag for an arbitrarily shaped particle:

$$\frac{F_{\text{Oseen}}}{F_{\text{Stokes}}} = 1 + \frac{F_{\text{Stokes}} Re_p}{16 \pi \mu_f U_p R} \tag{3-13}$$

where R = characteristic particle dimension.

When a particle is flowing near a wall, the drag is profoundly influenced. For the settling of a sphere of radius R along the axis of a circular cylinder of radius R_{CY}, Ladenburg (1907) obtained

$$\frac{F_{\text{bounded}}}{F_{\text{Stokes}}} = 1 + \frac{2.1044 R}{R_{CY}} \tag{3-14}$$

Brenner (1962) has also studied the flow of a particle near a wall and suggests the following format:

$$\frac{F_{\text{bounded}}}{F_{\text{Stokes}}} = \frac{1 - (U_f - U_p)/U_f}{1 - k F_{\text{Stokes}}/(6 \pi \mu_f U_p R)} \tag{3-15}$$

where U_f = fluid velocity
U_p = particle velocity
k = constant depending on the bounding wall

When there is movement along the axis of a circular cylinder (R distance of particle center to plane), $k = 2.1044$; when the particle is falling parallel to the plane, $k = \frac{9}{8}$; when the particle is falling midway between, $k = \frac{9}{16}$; and when the particle is falling parallel to two of the planes, $k = 1.004$. (Brodkey, 1967).*

In the case that the particle is so small that its size approaches the mean free path of the fluid, the Cunningham correction factor can be used. If l is the mean free path of the particle and D_p is its diameter, the drag relations can be given as

$$\frac{F_{\text{Cunningham}}}{F_{\text{Stokes}}} = \frac{1}{1 + 2A^*l/D_p} \tag{3-16}$$

where $A^* \approx 1$.

In addition to the particle diameter and mean free path comparison, Brownian movement is of importance for particle sizes less than 3 μm (Brodkey, 1967).

Example 3-4 Using the conditions on a 100-μm particle as seen in Example 3-3, part (c), compute the value of the terminal velocity change if Oseen's correction is used for the drag coefficient.

Substituting Eq. 3-12 into Eq. 3-7 yields a quadratic equation in U_p:

$$\left(\frac{3D_p\rho_f}{16\mu_f}\right)U_p^2 + U_p - \frac{(\rho_p - \rho_f)gD_p^2}{18\mu_f} = 0$$

Solving by standard techniques shows

$$U_p = 0.131 \text{ m/sec}$$

This is a sizable difference from the straight Stokes equation, which gave

$$U_p = U_t = 0.205 \text{ m/sec}$$

3-5 PARTICLE ROTATION

Particles will tend to rotate in a shear field. These rotation factors are normally of secondary importance. Impacts of the particles with a pipe wall, shear with a fluid eddy, and impacts between particles all can contribute to rotation of particles. In turbulent flow this rotation is the only factor of significance near the wall because of the boundedness of this region, as opposed to the more random nature of the center core of the pipe.

If a particle is assumed to be a small fluid element in a fluid with a velocity gradient dU/dy, the angular rotation $\bar{w}$ is given as

$$\bar{w} = \frac{1}{2}\frac{dU}{dy} \tag{3-17}$$

When a particle is rotating in a stationary fluid, a torque develops due to the fluid viscosity,

*Reproduced by permission of R. Brodkey.

$$T' = \pi\mu_f D_p^3 \bar{w} \tag{3-18}$$

This rotation and/or shear rate can cause a transverse force to be placed on the particle; this force is known as the Magnus force F_L and it gives a lift force of order of magnitude of $\rho_f U_L D_p^3 \, dU_t/dy$. In an analytical analysis of such systems, the non-linear terms of the Navier–Stokes generate this lateral velocity U_L. Generally, however, as stated before, the ratio of the left force to the Stokes force is much less than one. The Magnus force is only sizable for particles greater than 100 µm diameter.

3-6 CLOUDS

Clouds of particles often occur in industrial situations. Fluidized beds, discharges from bunkers, and spray-dryers are a few units where clouds of particles exist. It is possible to treat these clouds as entities in themselves and as such their behavior can be predicted.

A cloud has the ability to entrain the surrounding gas. The finer the particles in the cloud, the more readily this can be achieved. A cloud is assumed to have a uniform concentration, a slip velocity with the gas, and a definable diameter (Boothroyd, 1971). The terminal velocity of a cloud can be given as

$$U_{t(\text{cloud})} = \left[\frac{4(\rho_{ds} - \rho_f)D_c g}{3C_{D(\text{cloud})}\rho_f}\right]^{1/2} \tag{3-19}$$

where ρ_{ds} = density of the cloud
D_c = diameter of the cloud

Internally, the particles of the cloud have a smaller velocity,

$$U_{p(\text{internal})} = \left(\frac{4\rho_p g D_p}{3\rho_f C_{D(\text{particle})}}\right)^{1/2} \tag{3-20}*$$

The cloud velocity for a coal sample is determined in Example 3-5.

Example 3-5 A cloud of coal particles is traveling through a system. The cloud diameter is estimated to be 0.0254 m and the individual particles are 60 µm in diameter with a density of 1200 kg/m³. The total cloud is moving such that its drag coefficient is constant at 0.44. The individual particles in the cloud move much slower, such that the drag coefficient is given by Stokes' law. The system has atmospheric air transporting the solids, and the cloud density is 160 kg/m³. Determine the ratio of internal velocity to cloud particle velocity.

From Eq. 3-19 the cloud velocity is

$$U_{t(\text{cloud})} = \left[\frac{4(\rho_{ds} - \rho_f)D_c g}{3\rho_f C_{D(\text{cloud})}}\right]^{1/2}$$

*Reproduced by permission of Taylor & Francis Ltd., Boothroyd, 1971;

With $\rho_{ds} = 160 \text{ kg/m}^3$
$\rho_f = 1.2 \text{ kg/m}^3$
$D_c = 0.0254 \text{ m}$
$C_{D(\text{cloud})} = 0.44$

one has

$$U_{t(\text{cloud})} = 9.99 \text{ m/sec}$$

For the individual particles, Eq. 3-20 gives

$$U_{p(\text{internal})} = \left(\frac{\rho_p g D_p^2}{18 \mu_f} U_{p_i} \right)^{1/2} \qquad \text{where} \quad C_{D(\text{particle})} = \frac{24}{Re_p}$$

With $D_p = 60 \ \mu\text{m}$
$\mu_f = 0.01 \text{ cP}$

the equation gives

$$U_{p(\text{internal})} = 0.235 \text{ m/sec}$$

Therefore,

$$\frac{U_{p(\text{internal})}}{U_{t(\text{cloud})}} = 2.35 \times 10^{-2}$$

In his treatment of clouds of particles, Soo (1967) considered most systems having intimate contact all the way from packed beds through fluidization to dense-phase pneumatic transport. His analysis of clouds by individual particle bombardments between themselves and the confining wall relies on impact mechanics. This approach has served as a basis for both heat and electrostatic charge transfer in particulate systems. The elasticity of the particles and contact times became vital parameters in the analysis.

Clouds may also be analyzed by use of the approach of Neale and Nader (1974), which will be discussed later in reference to the effect of a system of particles acting on one another.

3-7 TRAJECTORIES

Finding the path of the particle lends much information to the flow behavior of gas-solid systems. Very few measurement techniques can deal with the individual particle trajectory and its velocity. Some very important technological questions about particle flow may be made with such measurements. Laser Doppler velocimeters (LDV) are available to provide such experimental data. Numerical simulation offers another path for such analysis. The different aspect about finding the path of the particle is that the basic equation defining such behavior is nonlinear, requiring a numerical solution of the equations. This procedure should not be considered as too much of an obstacle, and one should be able to apply knowledge of numerical analysis and experience to such a problem. The basic equation or definition relates the particle position to its velocity and time as

$$x = \int_{t_0}^{t} u_p \, dt' + x_0 \tag{3-21}$$

$$y = \int_{t_0}^{t} v_p \, dt' + y_0 \tag{3-22}$$

The velocities can be found by integration of the equation in its entirety or by using some simplification. For the case of particle acceleration and drag force, one has

$$\left(\frac{\pi D_p^3}{6}\right) \rho_p \frac{du_p}{dt} = C_D \rho_f \frac{\pi D_p^2}{2} (u_f - u_p) \, [(u_f - u_p)^2 + (v_f - v_p)^2]^{1/2} \tag{3-23}$$

for the x direction and, for the y direction:

$$\left(\frac{\pi D_p^3}{6}\right) \rho_p \frac{dv_p}{dt} = C_D \rho_f \frac{\pi D_p^2}{2} (v_f - v_p) \, [(u_f - u_p)^2 + (v_f - v_p)^2]^{1/2} \tag{3-24}$$

In reality these equations are simplified forms of Eq. 3-1. They can be further simplified by relationships between the drag coefficient and the Reynolds number.

Besides the rectilinear motion of particles, the trajectories in vortex motion are of importance in cyclones and combustors. Consider Fig. 3-4, which describes a particle in a field having radial (**L**) and transverse (**M**) components of velocity and acceleration. In this diagram,

$$\frac{\mathbf{L'} - \mathbf{L}}{\Delta\theta} = \frac{\Delta\mathbf{L}}{\Delta\theta} \tag{3-25}$$

and

$$\lim_{\Delta\theta \to 0} \frac{\Delta\mathbf{L}}{\Delta\theta} = \lim_{\Delta\theta \to 0} \frac{\sin \Delta\theta/2}{\Delta\theta/2} = 1 = \mathbf{M} = \frac{d\mathbf{L}}{d\theta} \tag{3-26}$$

or

$$\frac{d\mathbf{M}}{d\theta} = -\mathbf{L}$$

Taking the time derivatives, one has

$$\frac{d\mathbf{L}}{dt} = \frac{d\theta}{dt} \mathbf{M} \quad \text{and} \quad \frac{d\mathbf{M}}{dt} = -\frac{d\theta}{dt} \mathbf{L} \tag{3-27}$$

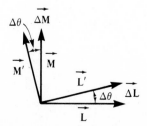

Figure 3-4 Vector diagram of forces.

The velocity is defined as

$$V = \frac{d\mathbf{R}}{dt} = \frac{d}{dt}(r\mathbf{L}) = \frac{dr}{dt}\mathbf{L} + r\frac{d\mathbf{L}}{dt} \tag{3-28}$$

or

$$V = \frac{dr}{dt}\mathbf{L} + \mathbf{M}r\frac{d\theta}{dt} \tag{3-29}$$

The second differentiation will yield the acceleration,

$$A = \frac{d^2r}{dt^2}\mathbf{L} + \frac{dr}{dt}\mathbf{L} + \frac{d^2\theta}{dt^2}\mathbf{M}r + \mathbf{M}\frac{d\theta}{dt}\frac{dr}{dt} + \frac{d\mathbf{M}}{dt}\frac{d\theta}{dt}r \tag{3-30}$$

or

$$A = \left[\frac{d^2r}{dt^2} - r\left(\frac{d\theta}{dt}\right)^2\right]\mathbf{L} + \left(r\frac{d^2\theta}{dt^2} + 2\frac{dr}{dt}\frac{d\theta}{dt}\right)\mathbf{M} \tag{3-31}$$

$$\underbrace{\qquad\qquad\qquad}_{\substack{\text{radial}\\\text{component}}} \qquad \underbrace{\qquad\qquad\qquad}_{\substack{\text{transverse}\\\text{component}}}$$

The mass times the acceleration terms can now be equated to any external forces acting on the system or particle.

Since the velocity components can be written as

$$v_r = \frac{dr}{dt} \tag{3-32}$$

$$v_t = r\frac{d\theta}{dt} \tag{3-33}$$

one can express Newton's second law for the radial and transverse directions as

$$m_p\left(\frac{dv_r}{dt} - \frac{v_t^2}{r}\right) = F_r \tag{3-34}$$

$$m_p\left(\frac{dv_t}{dt} + \frac{2v_r v_t}{r}\right) = F_t \tag{3-35}$$

In these expressions, F_r and F_t are drag and external forces acting on the particle in the radial and transverse directions, respectively. The external forces may be due to centrifugal, magnetic, or electrical interactions.

For the centrifugal case,

$$F_r = -6\pi R\mu_f v_r - \frac{\pi}{6}D_p^3\rho_f\frac{v_t^2}{r} \tag{3-36}$$

$$\underbrace{\qquad\qquad}_{\text{drag force}} \quad \underbrace{\qquad\qquad}_{\substack{\text{centrifugal}\\\text{force}}}$$

where u_t = transverse velocity of the fluid.

Thus, one can write the total balance as

$$m_p\left(\frac{dv_r}{dt} - \frac{v_t^2}{r}\right) = -6\pi R\mu_f v_r - \frac{\pi}{6}D_p^3\,\rho_f\,\frac{v_t^2}{r} \tag{3-37}$$

for the Stokes drag region. In this domain the term dv_r/dt is often small in comparison to the other terms. Assuming the velocity of the particle and fluid has no slip in the transverse direction, Eq. 3-37 reduces to

$$v_r = v_t^2\,\frac{(\rho_p - \rho_f)D_p^2}{18\mu_f r} \tag{3-38}$$

Application of these equations for a cyclone is given in Example 3-6.

Example 3-6 In a cyclone, assume that Stokes' law applies to a 10-μm-size particle having a density of 2700 kg/m^3. Air is flowing in the cyclone at STP with a volumetric rate of 0.0197 m^3/sec. Assume that the centrifugal force component dominates and that the slip velocity is low. The tangential velocity can be given as

$$U_t r = U_T\,(\text{constant}) = 0.323 \text{ m}^2/\text{sec}$$

If the inner radius of the cyclone is 0.011 m and the outer measurement is 0.0254 m diameter, how long will it take a 10-μm particle to drift from the smaller radius to the larger radius?

Employing Eq. 3-38, one has

$$v_r = \frac{v_t^2\,(\rho_p - \rho_f)D_p^2}{18\mu_f r}$$

Now,

$$v_r = \frac{dr}{dt}$$

and

$$v_t r = U_T = 0.323 \text{ m}^2/\text{sec}$$

$$r^2\,\frac{dr}{dt} = \frac{U_T^2\,(\rho_p - \rho_f)D_p^2}{18\mu_f}$$

$$\int_{R_1}^{R_2} r^3\,dr = \frac{U_T^2(\rho_p - \rho_f)D_p^2}{18\mu_f}\int_0^t dt$$

$$t = \frac{9}{2}\mu_f(R_2^4 - R_1^4)\,\frac{1}{U_T^2(\Delta\rho)D_p^2}$$

Utilizing the data,

$$t = \frac{(\tfrac{9}{2}\times 10^{-5} \text{ kg/m}\cdot\text{sec})\,[(0.0254)^4 - (0.011)^4]\text{ m}^4}{[(0.323)^2 \text{ m}^4/\text{sec}^2]\times[(2700 - 1.2)\text{ kg/m}^3]\times[(10\times 10^{-6})^2 \text{ m}^2]}$$

$$= 6.42\times 10^{-4} \text{ sec}$$

3-8 TWO OR MORE PARTICLES

After analyzing single-particle behavior in a fluid stream, it is natural to turn to two particles and then to multiple-particle interactions in a fluid. This last case is indeed complicated, and to date one must rely on empirical or semiempirical analysis of such systems. However, the study of doublets and triplets can add to the information base on a theoretical scale that may lead to analysis of more complicated systems in the future.

The earliest analysis of the motion of two rigid spheres began with Jeffery (1915). He considered the motion of two rigid spheres that rotated about their centers. Later, Stimson and Jeffery (1926) solved the axisymmetric problem of two spheres translating at an equal velocity along their lines of center. This exact solution serves as a reference as to the accuracy of other approximate treatments. The solution is based on determining Stokes' stream function for the motion of the fluid and from this the forces necessary to maintain the motion of the spheres. The force for equal spheres is given as

$$F = 3\pi\mu_f D_p U\lambda \tag{3-39}$$

with the correction to the Stokes equation λ given as

$$\lambda = \frac{4}{3}\sinh\alpha \sum_{n=1}^{\infty} \frac{n(n+1)}{(2n-1)(2n+3)}\left[1 - \frac{4\sinh^2(n+\frac{1}{2})\alpha - 2(n+1)^2\sinh^2\alpha}{2\sinh(2n+1)\alpha + (2n+1)\sinh 2\alpha}\right]$$

and the geometric factor $\alpha = 0$ for spheres touching and $\alpha = \infty$ for single spheres.

O'Neill (1970) and coworkers solved the asymmetrical motion of two equal spheres and the motion of two spheres approaching each other or a solid wall. With these works and that of Goldman, Cox, and Brenner (1967), a complete solution of forces acting on two equal solid psheres moving in an unbounded medium along their line of centers in known.

The comprehensive work of Happel and Brenner (1965) treats motion of two spheres as well as many particles. An approximate solution for the case of two spherical particles of the same diameter that are isotropic with respect to translation and rotation in an unbounded medium at rest at infinity gives the force as

$$\mathbf{F} = -3\pi\mu_f D_p\left[\frac{\mathbf{i}U_x}{1 + \frac{3}{4}(D_p/2\,\overline{l}) + \frac{1}{2}(D_p^3/8\overline{l}^3)} + \frac{\mathbf{k}U_z}{1 + \frac{3}{2}(D_p/2\,\overline{l}) - (D_p^3/8\overline{l}^3)}\right] \tag{3-40}$$

where $\mathbf{i}$ and $\mathbf{k}$ are unit vectors. In this expression, $\overline{l}$ is the distance between the centers of the spheres. Note that the term in brackets can be considered as a modification of the Stokes' equation for a single particle. The parameter λ in Eq. 3-39 is a convenient measure of the drag force exerted on each sphere when two spheres fall in an infinite viscous fluid compared to that exerted on a single sphere. For the case of spheres falling along the line of centers, the data of Bart (1959) taken at Reynolds numbers less than 0.05 with a sphere/cylinder diameter ratio of 0.0163 agrees well with Stimson and Jeffery's equation. The Happel and Pfeffer (1960) data show reasonable agreements for the sphere/cylinder diameter of 0.0437. Generally, for the case of spheres close together, the theory is substantiated by experiments (Evenson,

Hall, and Ward, 1959). Similar good agreement is seen between two spheres falling perpendicular to their lines of centers, as theoretically solved by Wakiya (1957) and from experimental data of Bart and Evenson, Hall, and Ward.

A comparison is given by Happel and Brenner (1965) of the resistance coefficient λ by use of Eq. 3-39 (the exact solution). Eq. 3-40 (an approximate solution), and the data of Bart for one sphere above the other. The results are presented in Table 3-2.

Table 3-2 Resistance coefficients, λ

Method of solution	λ
Exact, Eq. 3-39	0.645
Approximate, Eq. 3-40	0.616
Bart's data	0.647

(Happel and Brenner, 1965).
Reproduced by permission of authors.

For the case of several spheres, Burgers (1940) made calculations based on a first reflection technique for assemblages consisting of a small number of spheres held rigidly in position with respect to each other. The reflection technique involves a successive iteration technique where the boundary conditions are solved to any degree of approximation by considering the boundary conditions associated with one particle. Burgers found the drag on a system of particles to be

$$F = 3\pi\mu_f D_p U C_D \tag{3-41}$$

For two spheres in a line,

$$C_D = \frac{2}{1 + (D_p/2\bar{l})} \tag{3-42}$$

For three spheres in a line,

$$C_D = \frac{3}{1 + \frac{19}{3}(D_p/2\bar{l}) - \frac{1}{4}(D_p/2\bar{l})^2} \tag{3-43}$$

In order to account for the change in drag on a single particle by other particles in their vicinity, Neale and Nader (1974) used a geometric model that is an extension of Brinkman's model, which looks at a reference sphere embedded in a porous mass. By considering incompressible flow, creeping fluid flow, and a homogeneous swarm of spherical particles, Neale and Nader solved this system of equations to calculate the ratio of the force on a containing cylinder of N particles calculated by Stokes' law to that using their results. This ratio is given as

$$\frac{N \cdot F_{\text{Stokes}}}{F_{\text{total}}} = \frac{9(1 - \epsilon)}{2(\alpha\epsilon)^2} \tag{3-44}$$

The values of the porosity ϵ and α, given as $R/\sqrt{k}$ where R is the particle radius and k is the permeability of the porous medium, are interrelated. This relationship is shown in Fig. 3-5.

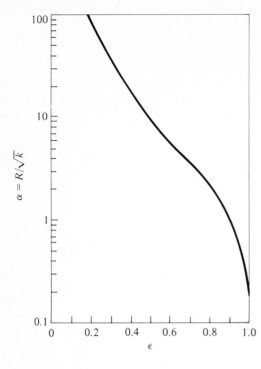

Figure 3-5 Dependence of α on ϵ for monodispersed spheres (Neale and Nador, 1974). Reproduced by permission of AIChE Journal.

3-9 TURBULENT SYSTEMS

To date, the treatments on single and multiple particles have not considered the most common condition of turbulent flow. The notable exception is the work of Torobin and Gauvin (1961). The basic turbulent flow equations for a system of particles can be written down, but their exact solution is unknown and remains one of the most challenging problems in fluid mechanics. The solution of the equations has the potential to give the velocities and concentrations in pneumatic transport lines. Certain information can be extracted from an analysis of the system of equations in general terms and that will be the goal of the approach given here.

The basic equations of fluid dynamics, consisting of the continuity plus the momentum analysis, are generally treated in detail for the slow laminar flow situations of single-phase and multiphase flow. When one is concerned with turbulent flow, the same general equations are valid by replacing the variables involved by a steady and fluctuating component of the variables. These expressions are valid, but they create a system of equations that cannot be solved because the number of variables outnumber the number of equations. Thus, a closure problem does exist. What can be done is to partially integrate the resulting equations and obtain some general formulas that may be interpreted qualitatively, or to employ simple analytical models suggested by experimental data that has been taken. This latter approach will be developed here starting with the single-phase continuity equation and continuing to the multiphase momentum expressions.

3-10 CONTINUITY EQUATIONS

The following development will be carried out for a cylindrical geometry paralleling the pipe flow situation. The density of the mixture of gas and solid can be given as

$$\rho_m = \sum_{q=1}^{n} \rho_q^* \tag{3-45}$$

where ρ_q^* is the density of component q based on the total volume of the mixture (Peters, 1970). Thus,

$$\rho_q^* = \frac{m_q}{V} \tag{3-46}$$

where m_q is the mass of component q and V is the entire volume occupied by the mixture. Using this basis, the continuity equation for component q in a cylindrical coordinate system is written as

$$\frac{\partial \rho_q^*}{\partial t} + \frac{1}{r} \frac{\partial (r \rho_q^* v_q)}{\partial r} + \frac{1}{r} \frac{\partial (\rho_q^* w_q)}{\partial x} = 0 \tag{3-47}$$

The velocities v_q, w_q, and u_q are the components of the velocities in the r, θ, and x directions, respectively.

Considering particulate flow, the particle number density concept may be employed, showing

$$\rho_q^* = m_q N_q \tag{3-48}$$

where m_q is the mass of a single particle and N_q is the particle number density of component q. Using this number density expression, one can rewrite the continuity equation as

$$\frac{\partial N_q}{\partial t} + \frac{1}{r} \frac{\partial (r N_q v_q)}{\partial r} + \frac{1}{r} \frac{\partial (N_q w_q)}{\partial \theta} + \frac{\partial (N_q u_q)}{\partial x} = 0 \tag{3-49}$$

The above development is a general presentation for multicomponent systems. Considering only two components, a gas and one solid, q may be represented as p for the solid and N_q as N in the following. In addition, for the gas phase the subscripts are dropped. In the turbulent situation, the number density of the solids varies. This variation is assumed to be irregular about a mean value. Thus, one can write

$$N = \bar{N} + N'$$

$$v_p = \bar{v}_p + v_p'$$

$$w_p = \bar{w}_p + w_p' \tag{3-50}$$

$$u_p = \bar{u}_p + u_p'$$

Substituting these expressions in Eq. 3-49 and time averaging gives

$$\frac{\partial \bar{N}}{\partial t} + \frac{1}{r} \frac{\partial (r\bar{N}\bar{v}_p)}{\partial r} + \frac{1}{r} \frac{\partial (\bar{N}\bar{w}_p)}{\partial \theta} + \frac{\partial (\bar{N}\bar{u}_p)}{\partial x}$$

$$\overbrace{\phantom{+ \frac{1}{r} \frac{\partial (r\overline{N'v'_p})}{\partial r} + \frac{1}{r} \frac{\partial (\overline{N'w'_p})}{\partial \theta} + \frac{\partial (\overline{N'u'_p})}{\partial x}}}^{\text{terms that result from fluctuations in } N}$$

$$+ \frac{1}{r} \frac{\partial (r\overline{N'v'_p})}{\partial r} + \frac{1}{r} \frac{\partial (\overline{N'w'_p})}{\partial \theta} + \frac{\partial (\overline{N'u'_p})}{\partial x} = 0 \qquad (3\text{-}51)$$

Assuming no θ dependence and homogeneous turbulence in the axial direction, $\partial \overline{N'u'_p}/\partial x = 0$, for a steady state, one has $\partial \bar{N}/\partial t = 0$. Thus, Eq. 3-51 reduces to

$$\frac{\partial (r\bar{N}\bar{v}_p)}{\partial r} + \frac{\partial (r\overline{N'v'_p})}{\partial r} = 0 \qquad (3\text{-}52)$$

This equation can be partially integrated from 0 to r to give

$$\bar{N}\bar{v}_p = -\overline{N'v'_p} + \frac{C(x)}{r} \qquad (3\text{-}53)$$

where $C(x)$ is a function depending on x. At the radius of the tube ($r = 0$), $\bar{v}_p$ is finite, thus, $C(x) = 0$, giving

$$\bar{N}\bar{v}_p + \overline{N'v'_p} = 0 \qquad (3\text{-}54)$$

Some time-average quantities are given in Example 3-7.

Example 3-7 Consider the steady-state continuity equation for an incompressible fluid in cylindrical coordinates. Using the turbulence representations for the velocities, determine the time-average contributions to this representation.

The continuity equation of concern is

$$\frac{1}{r} \frac{\partial}{\partial r} rv + \frac{1}{r} \frac{\partial w}{\partial \theta} + \frac{\partial u}{\partial x} = 0$$

Using

$$v = \bar{v} + v'$$

$$w = \bar{w} + w'$$

$$u = \bar{u} + u'$$

and substituting in the above, yields

$$\frac{1}{r} \frac{\partial}{\partial r} (r\bar{v} + rv') + \frac{1}{r} \frac{\partial}{\partial \theta} (\bar{w} + w') + \frac{\partial}{\partial x} (\bar{u} + u') = 0$$

Integrating this expression with respect to t and 0 to T and dividing by the time interval T, one has

$$\frac{1}{T}\int_0^T \frac{1}{r}\frac{\partial}{\partial r}\,r\bar{v}\,dt + \frac{1}{T}\int_0^T \frac{1}{r}\frac{\partial rv'}{\partial r}\,dt + \frac{1}{T}\int_0^T \frac{1}{r}\frac{\partial \bar{w}}{\partial \theta}\,dt$$

$$+ \frac{1}{T}\int_0^T \frac{1}{r}\frac{\partial w'}{\partial \theta}\,dt + \frac{1}{T}\int_0^T \frac{\partial \bar{u}}{\partial x}\,dt + \frac{1}{T}\int_0^T \frac{\partial u'}{\partial x}\,dt = 0$$

This expression reduces to

$$\frac{1}{r}\frac{\overline{\partial r\bar{v}}}{\partial r} + \frac{1}{r}\frac{\overline{\partial (rv')}}{\partial r} + \frac{1}{r}\frac{\overline{\partial \bar{w}}}{\partial \theta} + \frac{1}{r}\frac{\overline{\partial w'}}{\partial \theta} + \overline{\frac{\partial \bar{u}}{\partial x}} + \overline{\frac{\partial u'}{\partial x}} = 0$$

Since the time average of the fluctuating components is zero, the second, fourth, and sixth terms are zero. Thus, one has

$$\frac{1}{r}\frac{\overline{\partial r\bar{v}}}{\partial r} + \frac{1}{r}\frac{\partial \bar{w}}{\partial \theta} + \frac{\partial \bar{u}}{\partial x} = 0$$

3-11 EQUATIONS OF MOTION

Upon introduction of particles to a fluid phase, interaction terms must be incorporated into the single fluid momentum equations. These interaction terms are expressed as drag forces. The equations of motion in three directions can be written in the cylindrical system for an incompressible, Newtonian fluid:

$$\rho\left(\frac{\partial v}{\partial t} + v\frac{\partial v}{\partial r} + \frac{w}{r}\frac{\partial v}{\partial \theta} - \frac{w^2}{r} + u\frac{\partial v}{\partial x}\right) = -\frac{\partial p}{\partial r}$$

$$+ \mu\left[\frac{\partial}{\partial r}\frac{-1}{r}\frac{\partial (rv)}{\partial r} + \frac{1}{r^2}\frac{\partial^2 v}{\partial \theta^2} - \frac{2}{r^2}\frac{\partial w}{\partial \theta} + \frac{\partial^2 v}{\partial x^2}\right] + \rho g_r + \frac{m_p N}{\tau_p}(v_p - v) \quad (3\text{-}55)$$

$$\rho\left(\frac{\partial w}{\partial t} + \frac{\partial w}{\partial r} + \frac{w}{r}\frac{\partial w}{\partial \theta} + \frac{vw}{r} + u\frac{\partial w}{\partial x}\right) = -\frac{1}{r}\frac{\partial p}{\partial \theta}$$

$$+ \mu\left[\frac{\partial}{\partial r}\frac{1}{r}\frac{\partial (rw)}{\partial r} + \frac{1}{r^2}\frac{\partial^2 w}{\partial \theta^2} + \frac{2}{r^2}\frac{\partial v}{\partial \theta} + \frac{\partial^2 w}{\partial \theta^2}\right] + \rho g_\theta + \frac{m_p N}{\tau_p}(w_p - w) \quad (3\text{-}56)$$

$$\rho\left(\frac{\partial u}{\partial t} + v\frac{\partial u}{\partial r} + \frac{w}{r}\frac{\partial u}{\partial \theta} + u\frac{\partial u}{\partial x}\right) = -\frac{\partial p}{\partial x}$$

$$+ \mu\left[\frac{1}{r}\frac{\partial}{\partial r}\,r\frac{\partial u}{\partial r} + \frac{1}{r^2}\frac{\partial^2 u}{\partial \theta^2} + \frac{\partial^2 u}{\partial x^2}\right] + \rho g_x + \frac{m_p N}{\tau_p}(u_p - u) \quad (3\text{-}57)$$

where the relaxation time is given as

$$\tau_p = \frac{m_p}{6\pi a_p \mu_f} \quad (3\text{-}58)$$

for the Stokes regime.

The equations of motion for the solid phase then can be written ignoring the terms $C, D,$ and E in Eq. 3-1 to give

$$N\left(\frac{\partial v_p}{\partial t} + v_p \frac{\partial v_p}{\partial r} + \frac{w_p}{r}\frac{\partial v_p}{\partial \theta} - \frac{w_p^2}{r} + u_p \frac{\partial v_p}{\partial x}\right) = Ng_r + \frac{N}{\tau_p}(v - v_p) \qquad (3\text{-}59)$$

$$N\left(\frac{\partial w_p}{\partial t} + v_p \frac{\partial w_p}{\partial r} + \frac{w_g}{r}\frac{\partial w_p}{\partial \theta} + \frac{v_p w_p}{r} + u_p \frac{\partial w_p}{\partial x}\right) = Ng_\theta + \frac{N}{\tau_p}(w - w_p) \qquad (3\text{-}60)$$

$$N\left(\frac{\partial u_p}{\partial t} + v_p \frac{\partial u_p}{\partial r} + \frac{w_p}{r}\frac{\partial u_p}{\partial \theta} + u_p \frac{\partial u_p}{\partial x}\right) = Ng_x + \frac{N}{\tau_p}(u - u_p) \qquad (3\text{-}61)$$

Using the same time-averaging techniques as previously noted for the equations of motion for the fluid, one has

$$\rho\left(\frac{\partial \bar{v}}{\partial t} + \bar{v}\frac{\partial \bar{v}}{\partial r} + \frac{\bar{w}}{r}\frac{\partial \bar{v}}{\partial \theta} - \frac{\bar{w}^2}{r} + \bar{u}\frac{\partial \bar{v}}{\partial x}\right) = -\frac{\partial \bar{p}}{\partial r} + \mu\left[\frac{\partial}{\partial r}\frac{1}{r}\frac{\partial(r\bar{v})}{\partial r} + \frac{1}{r^2}\frac{\partial^2 \bar{v}}{\partial \theta^2} - \frac{\partial \bar{w}}{\partial \theta} + \frac{\partial^2 \bar{v}}{\partial x^2}\right]$$

$$+ \rho g_r + \frac{m_p \bar{N}}{\tau_p}(\bar{v}_p - \bar{v}) - \rho\ \underbrace{\frac{1}{r}\frac{\partial(\overline{rv'v'})}{\partial r} + \frac{1}{r}\frac{\partial(\overline{v'w'})}{\partial \theta} - \frac{\overline{w'w'}}{r} + \frac{\partial(\overline{u'v'})}{\partial x}}_{\text{turbulent flow}}$$

turbulent interactions,
particle-fluid

$$+ \frac{m_p}{\tau_p}\ \overline{(N'v'_p - N'v')} \qquad (3\text{-}62)$$

$$\rho\left(\frac{\partial \bar{w}}{\partial t} + \bar{v}\frac{\partial \bar{w}}{\partial r} + \frac{w}{r}\frac{\partial \bar{w}}{\partial \theta} + \frac{\bar{v}\bar{w}}{r} + \bar{u}\frac{\partial \bar{w}}{\partial x}\right) = -\frac{1}{r}\frac{\partial \bar{p}}{\partial \theta}$$

$$+ \mu\underbrace{\left[\frac{\partial}{\partial r}\frac{1}{r}\frac{\partial(r\bar{w})}{\partial r} + \frac{1}{r^2}\frac{\partial^2 \bar{w}}{\partial \theta^2} + \frac{2}{r^2}\frac{\partial \bar{v}}{\partial \theta} + \frac{\partial^2 \bar{w}}{\partial x^2}\right]}_{\text{turbulent flow}} + \rho g_\theta + \underbrace{\frac{m_p \bar{N}}{\tau_p}(\bar{w}_p - \bar{w})}_{\substack{\text{turbulent interactions,}\\ \text{particle-fluid}}}$$

$$- \rho\underbrace{\left[\frac{1}{r}\frac{\partial(\overline{rv'w'})}{\partial r} + \frac{1}{r}\frac{\partial(\overline{w'w'})}{\partial \theta} + \frac{\overline{v'w'}}{r} + \frac{\partial(\overline{u'w'})}{\partial x}\right]}_{} + \frac{m_p}{\tau_p}\overline{(N'w'_p - N'w')} \qquad (3\text{-}63)$$

$$\rho\left(\frac{\partial \bar{u}}{\partial t} + \bar{v}\frac{\partial \bar{u}}{\partial r} + \frac{w}{r}\frac{\partial \bar{u}}{\partial \theta} + \bar{u}\frac{\partial \bar{u}}{\partial x}\right) = -\frac{\partial \bar{p}}{\partial x} = \mu\underbrace{\left(\frac{1}{r}\frac{\partial}{\partial r}\ r\frac{\partial \bar{u}}{\partial r} + \frac{1}{r^2}\frac{\partial^2 \bar{u}}{\partial x^2}\right)}_{\text{turbulent flow}} + \rho g_x$$

turbulent interactions,
particle-fluid

$$+ \frac{m_p \bar{N}}{\tau_p}(\bar{u}_p - \bar{u}) - \rho\left[\frac{1}{r}\frac{\partial(\overline{rv'v'})}{\partial r} + \frac{1}{r}\frac{\partial(\overline{u'w'})}{\partial \theta} + \frac{\partial(\overline{u'u'})}{\partial x}\right]$$

turbulent interactions,
particle-fluid

$$+ \frac{m_p}{\tau_p}\overline{(N'u'_p - N'u')} \qquad (3\text{-}64)$$

For steady flow in the horizontal x direction alone, Eqs. 3-62, 3-63, and 3-64 can be simplified:

$$0 = \frac{m_p \bar{N} \bar{v}_p}{\rho \tau_p} - \frac{1}{r}\frac{\partial(\overline{rv'v'})}{\partial r} + \frac{\overline{w'w'}}{r} - \frac{\partial(\overline{u'v'})}{\partial x} + \frac{m_p}{\rho \tau_p}(\overline{N'v'_p} - \overline{N'v'}) \qquad (3\text{-}65)$$

$$0 = \frac{m_p \bar{N} \bar{w}_p}{\rho \tau_p} - \frac{1}{r}\frac{\partial(\overline{rv'w'})}{\partial r} - \frac{\overline{v'w'}}{r} - \frac{\partial(\overline{u'w'})}{\partial x} + \frac{m_p}{\rho \tau_p}(\overline{N'w'_p} - \overline{N'w'}) \qquad (3\text{-}66)$$

$$\frac{1}{\rho}\frac{\partial \bar{p}}{\partial x} = \frac{\nu}{r}\frac{\partial}{\partial r}\, r\frac{\partial \bar{u}}{\partial r} + \frac{m_p N}{\rho \tau_p}(\bar{u}_p - \bar{u}) - \frac{1}{r}\frac{\partial(\overline{ru'v'})}{\partial r} - \frac{\partial(\overline{u'u'})}{\partial x} + \frac{m_p}{\rho \tau_p}(\overline{N'u'_p} - \overline{N'u'})$$

$$(3\text{-}67)$$

Assuming no variation in the x direction for the terms $\overline{u'v'}$, $\overline{u'w'}$, and $\overline{u'u'}$, one has

$$0 = \frac{m_p \bar{N} \bar{v}_p}{\rho \tau_p} - \frac{1}{r}\frac{\partial(\overline{rv'v'})}{\partial r} + \frac{\overline{w'w'}}{r} + \frac{m_p}{\rho \tau_p}(\overline{N'v'_p} - \overline{N'v'}) \qquad (3\text{-}68)$$

$$\frac{1}{r}\frac{\partial(\overline{rv'w'})}{\partial r} = -\frac{\overline{v'w'}}{r} + \frac{m_p}{\rho \tau_p}(\overline{N'w'_p} - \overline{N'w'}) \qquad (3\text{-}69)$$

$$\frac{1}{\rho}\frac{\partial \bar{p}}{\partial x} = \frac{\nu}{r}\frac{\partial}{\partial r}\, r\frac{\partial \bar{u}}{\partial r} + \frac{m_p \bar{N}}{\rho \tau_p}(\bar{u}_p - \bar{u}) - \frac{1}{r}\frac{\partial(\overline{ru'v'})}{\partial r} + \frac{m_p}{\rho \tau_p}(\overline{N'u'_p} - \overline{N'u'})$$

$$(3\text{-}70)$$

The partial integration of Eq. 3-69 is shown in Example 3-8.

Example 3-8 Perform a partial integration on Eq. 3-9, the θ direction equation of motion of the fluid.

Equation 3-69 can be rearranged to give

$$\frac{\partial \overline{rv'w'}}{\overline{rv'w'}} = -\frac{\partial r}{r} + \frac{m_p}{\overline{v'w'}\rho \tau_p}(\overline{N'w'_p} - \overline{N'w'})\,\partial r$$

Integrating this from 0 to r, one has

$$\ln(r^2 \overline{v'w'}\,\big|_r - \ln(r^2 \overline{v'w'})\,\big|_0 = \int_0^r \frac{m_p}{\overline{v'w'}\rho \tau_p}(\overline{N'w'_p} - \overline{N'w'})\,dr$$

which can be written as

$$v'w' = \overline{v'w'}\,\big|_{r=0}\, \exp\left[\int_0^r \frac{m_p}{\overline{v'w'}\rho \tau_p}(\overline{N'w'_p} - \overline{N'w'})\,dr\right]$$

For the particulate phase one can write the equations of motion with time-averaging procedures as

$$\bar{N}\left(\frac{\partial \bar{v}_p}{\partial t} + \bar{v}_p \frac{\partial \bar{v}_p}{\partial r} + \frac{\bar{w}_p}{r}\frac{\partial \bar{v}_p}{\partial \theta} - \frac{\bar{w}_p^2}{r} + \bar{u}_p \frac{\partial \bar{v}_p}{\partial x}\right) = \bar{N}g_r + \frac{\bar{N}}{\tau_p}(\bar{v} - \bar{v}_p)$$

$$\left.\begin{aligned}
&- \frac{\partial(\overline{N'v_p'})}{\partial t} + \overline{N'v_p'}\frac{\partial \bar{v}_p}{\partial r} + \frac{\overline{N'w_p'}}{r}\frac{\partial \bar{v}_p}{\partial \theta} - \frac{\overline{N'w_p'}\bar{w}_p}{r} + \overline{N'u_p'}\frac{\partial \bar{v}_p}{\partial x} \\[6pt]
&- \frac{1}{r}\frac{\partial}{\partial r}\left(r\bar{N}\overline{v_p'v_p'} + r\overline{N'v_p'}\bar{v}_p + r\overline{N'v_p'v_p'}\right) \\[6pt]
&- \frac{1}{r}\frac{\partial}{\partial \theta}\left(\bar{N}\overline{v_p'w_p'} + \overline{N'v_p'}\bar{w}_p + \overline{N'v_p'w_p'}\right) \\[6pt]
&+ \frac{\bar{N}\overline{w_p'w_p'} + \overline{N'w_p'}\bar{w}_p + \overline{N'w_p'w_p'}}{r} \\[6pt]
&- \frac{\partial}{\partial x}\left(\bar{N}\overline{u_p'v_p'} + \overline{N'v_p'}\bar{u}_p + \overline{N'u_p'v_p'}\right) + \frac{1}{\tau_p}\left(\overline{N'v'} - \overline{N'v_p'}\right)
\end{aligned}\right\} \text{turbulent fluctuations} \qquad (3\text{-}71)$$

$$\bar{N}\left(\frac{\partial \bar{w}_p}{\partial t} + \bar{v}\frac{\partial \bar{w}_p}{\partial r} + \frac{\bar{w}_p}{r}\frac{\partial \bar{w}_p}{\partial \theta} + \frac{\bar{v}_L\bar{v}_p}{r} + \bar{u}_p\frac{\partial \bar{w}_p}{\partial x}\right) = \bar{N}g_\theta + \frac{\bar{N}}{\tau_p}(\bar{w} - \bar{w}_p)$$

$$\left.\begin{aligned}
&- \frac{\partial(\overline{N'w_p'})}{\partial t} + \overline{N'v_p'}\frac{\partial \bar{w}_p}{\partial r} + \frac{\overline{N'w_p'}}{r}\frac{\partial \bar{w}_p}{\partial \theta} + \frac{\overline{N'v_p'}\bar{w}_p}{r} + \overline{N'u_p'}\frac{\partial \bar{w}_p}{\partial x} \\[6pt]
&- \frac{1}{r}\frac{\partial}{\partial r}\left(r\bar{N}\overline{v_p'w_p'} + r\overline{N'w_p'}\bar{v}_p + r\overline{N'v_p'w_p'}\right) \\[6pt]
&- \frac{1}{r}\frac{\partial}{\partial \theta}\left(\bar{N}\overline{w_p'w_p'} + \overline{N'w_p'}\bar{w}_p + \overline{N'w_p'w_p'}\right) \\[6pt]
&- \frac{\bar{N}\overline{v_p'w_p'} + \overline{N'w_p'}\bar{v}_p + \overline{N'v_p'w_p'}}{r} \\[6pt]
&- \frac{\partial}{\partial x}\left(\bar{N}\overline{u_p'w_p'} + \overline{N'w_p'}\bar{u}_p + \overline{N'u_p'w_p'}\right) + \frac{1}{\tau_p}\left(\overline{N'w'} - \overline{N'w_p'}\right)
\end{aligned}\right\} \text{turbulent fluctuations} \qquad (3\text{-}72)$$

$$\bar{N}\left(\frac{\partial \bar{u}_p}{\partial t} + \bar{v}_p\frac{\partial \bar{u}_p}{\partial r} + \frac{\bar{w}_p}{r}\frac{\partial \bar{u}_p}{\partial \theta} + \bar{u}_p\frac{\partial \bar{u}_p}{\partial x}\right) = \bar{N}g_x + \frac{\bar{N}}{\tau_p}(\bar{u} - \bar{u}_p)$$

$$\left.\begin{aligned}
&- \frac{\partial(\overline{N'u_p'})}{\partial t} + \overline{N'v_p'}\frac{\partial \bar{u}_p}{\partial r} + \frac{\overline{N'w_p'}}{r}\frac{\partial \bar{u}_p}{\partial \theta} + \overline{N'u_p'}\frac{\partial \bar{u}_p}{\partial x} \\[6pt]
&- \frac{1}{r}\frac{\partial}{\partial r}\left(r\bar{N}\overline{u_p'v_p'} + r\overline{N'u_p'}\bar{v}_p + r\overline{N'u_p'v_p'}\right) \\[6pt]
&- \frac{1}{r}\frac{\partial}{\partial \theta}\left(\bar{N}\overline{u_p'w_p'} + \overline{N'u_p'}\bar{w}_p + \overline{N'w_p'u_p'}\right) \\[6pt]
&- \frac{\partial}{\partial \theta}\left(\bar{N}\overline{u_p'u_p'} + \overline{N'u_p'}\bar{u}_p + \overline{N'u_p'u_p'}\right) + \frac{1}{\tau_p}\left(\overline{N'u'} - \overline{N'u_p'}\right)
\end{aligned}\right\} \text{turbulent fluctuations} \qquad (3\text{-}73)$$

Additional simplifications can be made for the case of steady-state, homogeneous turbulence in the axial direction ($\partial/\partial x = 0$), no θ dependence, and $\bar{v} = \bar{w} = \bar{w}_p = 0$. Looking only at the equation for the z direction, one has the following:

$$0 = \frac{\bar{N}}{\tau_p}(\bar{u} - \bar{u}_p) - \frac{1}{r}\frac{\partial}{\partial r}(r\bar{N}\overline{u'_p v'_p} + r\overline{N' u'_p}\bar{v}_p + r\overline{N' u'_p v'_p})$$

$$+ \frac{1}{\tau_p}(\overline{N' u'} - \overline{N' u'_p}) \tag{3-74}$$

Comparing this with Eq. 3-70, one has

$$\frac{\partial \bar{p}}{\partial x} = \frac{1}{r}\frac{\partial}{\partial r}\left(r\mu \frac{\partial \bar{u}}{\partial r} - r\rho\overline{u'v'} - rm_p\bar{N}\overline{u'_p v'_p} - rm_p\overline{N' u'_p}\bar{v}_p - rm_p\overline{N' u'_p v'_p}\right) \tag{3-75}$$

This gives the time-averaged axial pressure gradient as a function of the fluid velocity profile, the radial particle velocity, and the turbulent fluctuations. This equation can be partially integrated, noting that the pressure gradient must be finite at $r = 0$:

$$\frac{\partial \bar{p}}{\partial x} = \frac{2}{r}\left(\mu_f \frac{\partial \bar{u}}{\partial r} - \rho\overline{u'v'} - m_p\bar{N}\overline{u'_p v'_p} - m_p\overline{N' u'_p}\bar{v}_p - m_p\overline{N' u'_p v'_p}\right) \tag{3-76}$$

In principle, all the quantities in Eq. 3-76 can be evaluated experimentally. If the assumption that N' is zero (i.e., the particulate phase density is constant) is made, then Eq. 3-76 simplifies to

$$\frac{\partial \bar{p}}{\partial x} = \frac{2}{r}\left(\mu_f \frac{\partial \bar{u}}{\partial r} - \rho\overline{u'v'} - m_p\bar{N}\overline{u'_p v'_p}\right) \tag{3-77}$$

It should be noted that Eq. 3-77 reduces to the special cases of (a) single-phase turbulent flow ($N = 0$) and (b) single-phase laminar flow ($N = u' = v' = 0$).

Additional analysis of these systems of turbulent gas-solid flow can be done by looking at these effects in reference to the single-phase results. If Eq. 3-77 is multiplied by $D/2\rho_m U^2$, then the friction factor results:

$$f_m = \frac{D/r}{\rho_m U^2}\left[\mu_f\left(\frac{\partial \bar{u}}{\partial r}\right)_m - \rho(\overline{u'v'})_m - m_p\bar{N}\overline{u'_p v'_p}\right] \tag{3-78}$$

The subscript m for the gas-phase velocity gradient and Reynolds stresses refers to these quantities when the particulate phase is present. Similarly, for the single-phase flow, one can perform the same operation and multiply by $D/2\rho U^2$ to obtain

$$f_g = \frac{D/r}{\rho U^2}\left(\mu_f \frac{\partial \bar{u}}{\partial r} - \rho\overline{u'v'}\right) \tag{3-79}$$

If both equations are applied at the same clean-air Reynolds number, Eq. 3-78 can be divided by Eq. 3-79 to obtain a useful relation:

$$\frac{f_m}{f_g} = \frac{\rho\left[\mu_f(\partial \bar{u}/\partial r)_m - \rho(\overline{u'v'})_m - m_p\bar{N}\overline{u'_p v'_p}\right]}{\rho_m\left[\mu_f(\partial \bar{u}/\partial r) - \rho\overline{u'v'}\right]} \tag{3-80}$$

The gas-phase velocity gradient and Reynolds stresses when the particles are present can be broken into two parts—those normally present at that clean-air Reynolds number and those contributions due to the presence of the particles. Equation 3-80 can now be written as

$$\frac{f_m}{f_g} = \frac{\rho}{\rho_m}\left[1 + \frac{\mu_f(\partial\bar{u}/\partial r)_\Delta - \rho(\overline{u'v'})_\Delta - m_p\overline{Nu_p'v_p'}}{\mu_f(\partial\bar{u}/\partial r) - \rho\overline{u'v'}}\right] \qquad (3\text{-}81)$$

The second term within the brackets in Eq. 3-81 is the ratio of the change in stresses due to the presence of the particles (subscripted Δ) divided by the stresses normally present at the particular clean-air Reynolds number. This term will be defined as Φ_T, the friction modification term. Finally, recognizing the definition of ρ_m at relatively dilute solids concentrations, Eq. 3-81 can be written as

$$\frac{f_m}{f_g} = \frac{1 + \Phi_T}{1 + (W_s/W_g)} \qquad (3\text{-}82)$$

where W_s is the solid flow rate and W_g is the gas flow rate.

Three cases for Φ_T can be considered. If $\Phi_T = 0$, then the turbulence characteristics are unchanged due to the particle addition. In this case,

$$\frac{f_m}{f_g} = \frac{1}{1 + (W_s/W_g)} \qquad (3\text{-}83)$$

and there is no change in pressure drop upon addition of the particles. For positive Φ_T, the combined stresses due to the velocity gradient, gas-phase Reynolds stresses, and particulate-phase Reynolds stresses are increased. If this situation is such that

$$\frac{1}{1 + (W_s/W_g)} < \frac{f_m}{f_g} < 1 \qquad (3\text{-}84)$$

then there is an increase in the pressure drop upon particle addition but not as great as predicted by the increased mass added to the flow. The final case is negative Φ_T, or

$$\frac{f_m}{f_g} < \frac{1}{1 + (W_s/W_g)} \qquad (3\text{-}85)$$

In this situation, the combined stresses are reduced and a pressure drop upon addition of solid particles to the flow is truly achieved (Peters and Klinzing, 1972).

PROBLEMS

3-1 Consider a 2-μm-diameter particle of fly ash with a specific gravity of 1.1 flowing in an acceleration region in air at standard conditions. If the acceleration of the fluid is twice the acceleration of the particle and is constant over a time period of 2 sec, compute the size of the Basset force on the particle. The particle acceleration is 5 m/sec^2. What occurs in a system under 4137 kN/m^2 pressure?

3-2 In a combustor a 100-μm-diameter particle of coal is accelerated from 0.305 m/sec to its final velocity. Assume that the drag force is the only dominant retardant on the particle and that the

Stokes drag range is applicable. Determine the time it takes the particle to accelerate to 95% of the fluid velocity in an atmospheric combustor. How does this time vary for a 1724-kN/m^2 pressurized unit?

3-3 For a 10-μm-diameter urania particle, determine the settling velocity in air at standard conditions. Compare this settling velocity to that of a similar size particle of iron oxide and coal.

3-4 A turbulent flow of air has a turbulence intensity of 30%. A 500-μm-diameter particle of limestone is flowing in this stream at STP. Determine the drag force on this particle. The particle slip velocity is 3.31 m/sec.

3-5 Ceramic particle of specific gravity = 2.5 are settling in stagnant air at standard conditions. The shape of the particles is not spherical but can be approximated as cubic and equilateral prisms. The characteristic length is 25 μm. Determine the settling velocities of these nonspherical particles.

3-6 Compute the drag force on a 50-μm-diameter coal particle settling near a planar wall in air at 101.3 kN/m^2 pressure and 38°C. The particle center is 2 diameters from the wall.

3-7 Determine the drag force on a particle 0.1 μm in diameter settling in air at standard conditions. The particle is sand.

3-8 For a swirling motion consider only the radial and transverse components of the velocity of a particle. Determine the velocities of a particle in such a field having only the drag force acting in resistance to the movement of the particle.

3-9 Determine the time-average equation for the r component of the fluid velocity using the technique shown in Example 3-7.

3-10 Combine the equation of motion for the fluid and particulate phase in the θ direction and integrate the result to obtain an expression for $v'w'$ in terms of the particle velocity fluctuations.

3-11 Interpret Fig. 4-13 in light of the expressions derived in Eqs. 3-82 to 3-85.

REFERENCES

Bart, E., MS.: Ch. E. Thesis, New York University, New York, 1959.

Boothroyd, R. G.: *Flowing Gas Solids Suspensions,* Chapman and Hall, London, 1971.

Brenner, H.: *J. Fluid Mech.* **11**:604 (1961).

——: *J. Fluid Mech.* **12**:35 (1962).

——: *Chem. Eng. Sci.* **18**:1 (1963); **19**:599 (1964); **21**:97 (1966).

Brodkey, R. S.: *The Phenomena of Fluid Motion,* Addison-Wesley, Reading, Mass., 1967.

Burgers, J. M.: *Proc. Koninge. Akad. Wetenschap (Amsterdam)* **43**:425 (1940).

Chao, B. T.: *Osterreich. Ing. Arkiv* **18**(1/2):7 (1964).

Claman, A., and Gauvin, W. H., *AIChE J.* **15**:184 (1969)

Clift, R., and W. H. Gauvin: *Chemeca* 14 (1970).

Corrsin, S., and J. Lumley: *Appl. Sci. Res.* **6A**:114 (1956).

Evenson, G. F., E. W. Hall, and S. G. Ward: *Brit. J. Appl. Phys.* **10**:43 (1959).

Goldman, A. J., R. G. Cox, and H. Brenner: *Chem. Eng. Sci.* **22**:637, 653 (1967).

Happel, J., and H. Brenner: *Low Reynolds Number Hydrodynamics,* Prentice-Hall, Englewood Cliffs, N.J., 1965.

—— and R. Pfeffer: *AIChE J.* **6**:129 (1960).

Hjelmfelt, A. T., Jr., and L. F. Mockros: *Appl. Sci. Res.* **A16**:149 (1966).

Jeffrey, G. B.: *Proc. London Math. Soc.* (Ser. 2) **14**:327 (1915).

Landenberg, R.: *Ann. Phys.* **23**:447 (1907).

McCabe, W. L., and J. C. Smith: *Unit Operations of Chemical Engineering,* 3d ed., McGraw-Hill, New York, 1976.

Neale, G. H., and W. K. Nader: *AIChE J.* **20**:530 (1974).

Odar, F.: *J. Fluid Mech.* **25**:591 (1966).

——: *Trans. ASME, J. Appl. Mech.* **35**:238, 652 (1968).

O'Neill, M. E.: *Appl. Sci. Res.* **21**:453 (1970).

Peters, L.: Ph.D. Thesis, *A Study of Two-phase, Solid-in-Air Flow Through Rigid and Compliant Wall Tubes,* University of Pittsburgh, 1970.

—— and G. E. Klinzing: *Can. J. Chem. Eng.* **50**:441 (1972).

Shapiro, A. H.: *Shape and Flow,* Anchor (Doubleday), New York, 1961.

Soo, S. L.: *Fluid Dynamics of Multiphase Systems,* Blaisdell, Waltham, Mass., 1967.

Stimson, M., and G. B. Jeffery: *Proc. R. Soc. (London)* **A111**:110 (1926).

Stokes, G. G.: *Trans. Cambridge Phil. Soc.* **9**:8 (1851).

Torobin, L. B., and W. H. Gauvin: *AIChE J.* **7**:615 (1961).

Tchen, C. M.: Thesis, Delft, Martenus Nijhoff, The Hague (1947).

Wadell, H.: *J. Franklin Inst.* **217**:459 (1934).

Wakiya, S.: *Niigata Univ. (Nagaoka, Japan) Coll. Eng. Res.* Rept. no. 6 (1957).

FOUR

GAS-SOLID PNEUMATIC TRANSFER

4-1 INTRODUCTION

In many respects Chaps. One to Three provide the foundation for designing piping systems for the transport of solids by a gas under a variety of conditions. With the advent of our concern for more energy, the increased production of coal is imperative, and large volumes of solid material must be moved at one or many stages in the conversion of coal to energy. Much work has been done on the movement of solids by gases, but many of the studies have been performed only because of specific needs or situations. Generally, empiricism has been involved in the designs and results are applicable only over limited ranges. To a great extent, one must still rely on some of this empiricism, but the field is now approaching a more comprehensive analysis and understanding of the gas-solid pneumatic flow systems. Yang (1976) provides a unified theory on dilute-phase pneumatic transfer. This study has done much to unify the dilute-phase systems, but questions still remain concerning the design of the dense choked-phase regions of flow and horizontal flow having saltation effects. These important regions of flow are finding increased use in industry.

4-2 DILUTE-PHASE TRANSPORT

Before proceeding with the dilute-phase development, an explanation of the difference between the dilute phase and dense phase is in order. If one considers the vertical transport of solids by a large quantity of gas, one finds a particular pressure drop. As the gas velocity is reduced at the same rate of transport of solids, the pressure drop decreases. This can be seen in Fig. 4-1 in going from point A to point B on the curve. At one particular velocity of the gas a minimum pressure drop is experienced. This minimum pressure drop point may be used as the dividing point between dilute- and dense-phase transfer. Lower gas velocities produce higher pressure drops and the

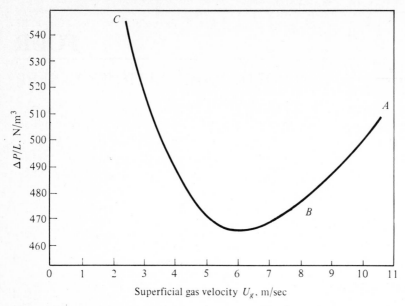

Figure 4-1 Pressure drop per unit length versus gas velocity (Konno and Saito, 1969).

choked-flow (slugging) conditions at point C. With increased pressure drop, a moving bed or extrusion flow will form, and this transition may cause solids to plug the line. A similar situation exists in the horizontal transport of solids, but with a more abrupt change in pressure drop. This change is due to saltation or settling at the bottom of the pipe, which causes a decrease in cross-sectional area of the pipe (see Fig. 4-10). Thus, in both vertical and horizontal transport, the region to the right of the minimum in the pressure-drop curve with gas-transport velocity will represent dilute-phase transfer.

4-3 PARTICLE FORCE BALANCE

Consider the flow of solids and gas as seen in Fig. 4-2. A force balance on the particles in the differential section dL may be represented as

$$\Delta m_s \quad \frac{dU_p}{dt} \quad = dF_d \quad - \quad dF_g \quad - \quad dF_f \qquad (4\text{-}1)$$

$$\text{mass} \qquad \text{acceleration} \qquad \text{drag} \quad \text{gravity} \qquad \text{friction}$$

for vertical transport, where Δm_s is the effective weight of the solid particles in the section. The gravity force $dF_g = 0$ for horizontal transport, and at a steady state, $dU_p/dt = 0$. The acceleration term is of importance when the particles are first injected into the gas stream and also when the particles undergo changes in flow direction, such as in pipe turns and bends. The drag force dF_d on the particles is the sum of the drag forces on each individual particle. This can be represented as

$$dF_d = \tfrac{3}{4} C_{Ds} \frac{\rho_f (U_f - U_p)^2}{(\rho_p - \rho_f) D_p} \, \Delta m_s \qquad (4\text{-}2)$$

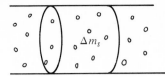

Figure 4-2 Differential balance diagram.

where U_f is the actual fluid velocity defined as the superficial gas velocity divided by the voidage. Yang (1976) has suggested modifying the individual drag coefficient C_{D_s} by the factor $\epsilon^{-4.7} C_{D_s} = C_{D_M}$, where ϵ is the voidage in the pipe flow. The voidage representing the volume of the pipe section occupied by the gas alone is defined as

$$\epsilon = 1 - \frac{4W_s}{(\rho_p - \rho_f)\pi D^2 U_p} \tag{4-3}$$

where W_s is the solid mass flow rate. The gravity force terms can be given as $g\,\Delta m_s$. The remaining force term is due to the solid friction loss. This force may be represented in terms of a friction factor:

$$dF_f = \frac{2f_s U_p^2}{D}\,\Delta m_s \tag{4-4}$$

Expansion of Eq. 4-1 yields

$$\frac{dU_p}{dt} = \tfrac{3}{4} C_{DM}\,\rho_f\,\frac{(U_f - U_p)^2}{(\rho_p - \rho_f)D_p} - g - \frac{2f_s U_p^2}{D} \tag{4-5}$$

At steady state, this equation can be solved for the value of U_p, which results in an implicit equation. The drag coefficient ratio at various voidages is given in Example 4-1.

Example 4-1 As noted in multiparticle flow systems, the drag coefficient is different from that for a single particle. Compute the effect of voidage on the multiparticle drag coefficient.

The multiple drag coefficient is given as

$$C_{D_m} = \epsilon^{-4.7} C_{D_s}$$

Consider the voidage of $0.99, 0.98, 0.95$, and 0.90 for a flow system. The first few values of high voidages are for dilute transport systems.

Table 4-1

C_{DM}/C_{D_s}	ϵ
1.048	0.99
1.10	0.98
1.27	0.95
1.64	0.90

4-4 OVERALL PRESSURE DROP

The total pressure drop in the pneumatic conveying line can be broken down into its individual contributions:

$$\Delta P_T = \Delta P_{acceleration} + \Delta P_{static} + \Delta P_{friction} \tag{4-6}$$

The static contribution due to the particles for vertical flow is

$$\Delta P_{s_p} = \rho_p(1 - \epsilon)L \cdot g \tag{4-7}$$

This term is zero for horizontal flow. The static factor due to the gas alone is generally much smaller than the solid contribution and is most often neglected. This factor may be written as

$$\Delta P_{s_g} = \rho_f \epsilon L \cdot g \tag{4-8}$$

The frictional contribution can also be divided into two parts: that due to the gas alone and that due to the solids. The gas frictional term is given as

$$\Delta P_{F_g} = \frac{2f_g \rho_f U_f^2 L}{D} \tag{4-9}$$

and the solids contribution is

$$\Delta P_{F_p} = \frac{2f_s \rho_p(1 - \epsilon)U_p^2 L}{D} \tag{4-10}$$

Care must be taken to note that some authors use a friction factor f_p in their analysis. The value of f_p is equal to $4f_s$.

The remaining term in the total pressure drop relation is the acceleration term. This contribution is essentially the similar static and frictional losses experienced by the solids, but acting over the acceleration length where the particle is being accelerated. The acceleration length may be found by the solution of Eq. 4-5, noting that $dL = U_p dt$, to give

$$L_{accel} = \int_{U_{p_1}}^{U_{p_2}} \frac{U_p \, dU_p}{\frac{3}{4}C_{D_s}\epsilon^{-4.7}\rho_f(U_f - U_p)^2/(\rho_p - \rho_f)D_p - (g + 2f_s U_p^2)/D} \tag{4-11}$$

for vertical transfer. For horizontal transfer, the contribution of gravity is equal to zero. The pressure drop over the acceleration length may thus be given as

$$\Delta P_{accel} = \int_0^{L_{accel}} \rho_p(1 - \epsilon)g \, dL + \int_0^{L_{accel}} \frac{2f_g \rho_f U_f^2 dL}{D}$$

$$+ \int_0^{L_{accel}} \frac{2f_s \rho_p(1 - \epsilon)U_p^2}{D} dL + \left[\rho_p(1 - \epsilon)U_p^2\right] \text{ at } L_{accel} \tag{4-12}$$

The upper and lower limits on Eq. 4-11 are of interest in the analysis of acceleration effects. The upper limit U_{p_2} is the value of the particle velocity at a final steady state. The initial value U_{p1} is the velocity from which the particle accelerates. This can be any value for which the system is designed.

The acceleration length of a vertical transport system using Eq. 4-11 is presented in Example 4-2.

Example 4-2 Consider the vertical transport of a Pittsburgh seam coal at a rate of 0.063 kg/sec in a 0.00107-m-ID pipe. The average particle size is 237 μm and its density is 1282 kg/m^3. Determine the acceleration length for the coal to reach its steady-state value. The gas velocity is 10.6 m/sec.

Using Eq. 4-11,

$$L_{accel} = \int_{U_{p1}}^{U_{p2}} \frac{U_p \, dU_p}{\frac{3}{4} C_{D_s} \epsilon^{-4.7} \rho_f (U_f - U_p)^2 / (\rho_p - \rho_f) D_p - (g + 2f_s U_p^2/D)}$$

In employing this equation, the solids friction factor is necessary. Several representations could be used for this quantity. Equation 4-15 for the dense-phase region employed, which permits this parameter to be a constant under the conditions of concern. The above expression may now be simplified with an average ϵ to

$$\Delta L = \int_{U_{p1}}^{U_{p2}} \frac{U_p \, dU_p}{\alpha_1 U_p^2 + \alpha_2 U_p + \alpha_3}$$

where
$$\alpha_1 = \frac{3}{4} \frac{C_{D_s} \epsilon^{-4.7} \rho_f}{D_p (\rho_p - \rho_f)} - \frac{2f_s}{D}$$

$$\alpha_2 = -\frac{3}{2} U_f \frac{C_{D_s} \epsilon^{-4.7} \rho_f}{D_p (\rho_p - \rho_f)}$$

$$\alpha_3 = \frac{3}{4} U_f^2 \frac{C_{D_s} \epsilon^{-4.7} \rho_f}{D_p (\rho_p - \rho_f)} - g$$

This expression can be integrated by standard formulas; letting

$$x = \alpha_1 U_p^2 + \alpha_2 U_p + \alpha_3$$

$$x = U_p$$

$$q = 4\alpha_1 \alpha_3 - \alpha_2^2$$

$$\int \frac{x \, dx}{X} = \frac{1}{2\alpha_1} \log X - \frac{\alpha_2}{2\alpha_1} \int \frac{dx}{X}$$

$$= \frac{1}{2\alpha_1} \log X - \frac{\alpha_2}{2\alpha_1} \frac{1}{\sqrt{-q}} \log \frac{2\alpha_1 x + \alpha_2 - \sqrt{-q}}{2\alpha_1 x + \alpha_2 + \sqrt{-q}}$$

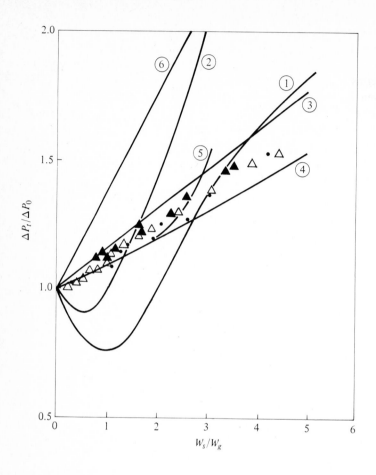

	Tube diameter, mm	Re	Particle	Particle size, μm	Gas	Reference
●	28	21,700	Copper	103	Air	
△	28	36,000	Copper	103	Air	Present work
▲	28	45,000	Copper	103	Air	
1	50.8	53,000	Zinc	10	Air	
2	76.2	53,000	Zinc	10	Air	
3	12.7	8,000	Graphite	5	Argon	Cited from Boothroyd (1966)
4	12.7	17,000	Graphite	5	Nitrogen	
5	18.0	27,400	Glass	36	Air	
6	–	45,000	–	–	–	Leung and Wiles (1976)

Figure 4-3 Frictional pressure drop ratio as a function of loading for the acceleration region. [Reprinted with permission from *Int. J. Multiphase Flow* **4**:53 (1978) by Shimizu, Echigo, and Hasegawa, Pergamon Press, Ltd.]

The drag coefficient can be found from the procedures outlined in Chap. Three to give a characteristic parameter (K) to determine the range of application for $q < 0$.

All the terms in the above expression are known except the voidage ϵ. The voidage is given by Eq. 4-3, however, the particle velocity is not known. An average value of ϵ is assumed for calculation ease. Using $U_f = U_p$ in Eq. 4-3 gives a value of ϵ that can be employed for analysis. This value should be rechecked after U_p is finally determined.

Substituting the appropriate values in the expression for acceleration length, the value of L_{accel} is calculated to be 0.567 m.

A study by Shimizu, Echigo, and Hasegawa (1978), showing an entry length for vertical flow of a gas-solid system to be monotonically increasing with a modified Reynolds number, is related to this acceleration length. These investigators define their modified Reynolds number as

$$Re_m = \left[\frac{(1-\epsilon)\rho_p U_p}{\epsilon \rho_f U_f} \right] \frac{\epsilon \rho_f U_f D}{\mu} \cong \left(1 + \frac{W_s}{W_g} \right) Re \qquad (4\text{-}13)$$

(see Fig. 4-3). The value of ΔP_t in Fig. 4-3 is the pressure drop seen with acceleration, while ΔP_0 is the base pressure drop without acceleration.

4-5 FRICTIONAL REPRESENTATION FOR SOLIDS PRESENCE

The equation presented for the representation of the pressure drops in piping systems for the transport of solids by gas seems to be quite straightforward. The one factor that is ill defined is that of the solid friction factor f_s. A number of investigators have attempted to analyze this friction factor, and most of the studies have been under the condition of dilute-phase transfer. Table 4-2 lists the various expressions for f_s suggested by several investigators. The easiest expression to use is Stemerding's; Yang gives an implicit representation. The question arises as to which representation for the friction factor is best for a particular application. Generally, in the dilute-phase region far removed from the minimum pressure drop point one can employ any of the expressions and obtain a fairly accurate representation of experimentally determined pressure drops. Figure 4-4 shows this behavior. Example 4-3 determines the pressure loss for a dilute coal transport system.

Example 4-3 A coal conversion pilot unit is using powdered coal at the rate of 0.0126 kg/sec. The transport line is 0.0254-m-ID steel pipe. The unit is designed so that it has roughly 9.1 m of vertical rise between the coal hopper and the entrance part to the process unit. The average particle size on a weight basis is 100 μm diameter. The transport gas at 6.89×10^3 N/m^2 of pressure is nitrogen with a linear velocity of 6.1 m/sec. Determine the energy loss for this vertical

Table 4-2 Solid friction factor for various models

Investigator	f_S
Stemerding (1962)	0.003
Reddy and Pei (1969)	$0.046\ U_p^{-1}$
Van Swaaij, Buurman, and van Breugel, (1970)	$0.080\ U_p^{-1}$
Capes and Nakamura (1973)	$0.048\ U_p^{-1.22}$
Konno and Saito (1969)	$0.0285\ \sqrt{gD}\ U_p^{-1}$
Yang (1978), vertical	$0.00315\ \dfrac{1-\epsilon}{\epsilon^3}\left[\dfrac{(1-\epsilon)U_t}{U_f-U_p}\right]^{-0.979}$
Yang (1976), horizontal	$0.0293\ \dfrac{1-\epsilon}{\epsilon^3}\left[\dfrac{(1-\epsilon)U_f}{\sqrt{gD}}\right]^{-1.15}$

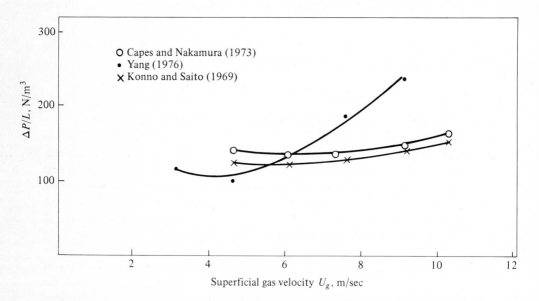

Figure 4-4 Model comparison of pressure drop per unit length as a function of superficial gas velocity.

transport scheme. Use the Konno–Saito expression for the particle friction factor and expression 1 of Table 4-3 for the particle velocity.

In analyzing this case the terminal velocity must first be calculated. Using the K criteria one finds

$$K = 3.50$$

and thus,

$$U_t = 0.347 \text{ m/sec}$$

Therefore,

$$U_p = U_g - U_t = 6.1 \text{ m/sec} - 0.347 \text{ m/sec} = 5.75 \text{ m/sec}$$

The frictional representation for the solids is that of Konno and Saito (1969):

$$f_s = 0.0285 \, U_p^{-1} \sqrt{gD}$$

$$\epsilon = 1 - \frac{4W_s}{\pi D_t^2 (\rho_p - \rho_f) U_p}$$

$$= 1 - \frac{4 \times 0.0126 \text{ kg/sec}}{\pi (0.0254)^2 \text{ m}^2 [(1200 - 1.2) \text{ kg/m}^3] \, 5.75 \text{ m/sec}}$$

$$= 0.996$$

The gas friction factor f_g is calculated from Koo's equation

$$f_g = 0.0014 + 0.125 \left(\frac{1}{Re^{0.32}} \right) = 0.0678$$

Thus, the overall pressure drop is then given as

$$\Delta P = \rho_p (1 - \epsilon) Lg + \rho_f \epsilon Lg + \frac{2 f_g \rho_f U_f^2 L}{D} + \frac{2 f_s \rho_p (1 - \epsilon) U_p^2 L}{D}$$

Substituting the values from above, gives

$$\frac{\Delta P}{L} = 113.7 \text{ N/m}^2/\text{m}$$

or, for 9.1 m length,

$$\Delta P = 1035 \, \frac{\text{N}}{\text{m}^2}$$

4-6 PARTICLE VELOCITY

The particle velocity is a very difficult parameter to determine experimentally. Pulse-injection techniques have been suggested for its measurement, as well as isolation of a section of the flow system by selenoid valves. The laser Doppler velocimeter system can be employed on dilute systems to measure the solids velocity. Table 4-3 gives a listing of expressions for the determination of U_p.

Expression 1 of Table 4-3 gives the particle velocity written in terms of the terminal velocity. This expression can be used generally for fine particles in the dilute-phase regime. Large particle sizes generally cause the particle velocity to be quite small and in some cases negative when inappropriate averaging has been done. Hinkle

Table 4-3 Particle velocity for models

Investigator	U_p
Expression 1	$U_g - U_t$
Hinkle (1953) IGT (1978)	$U_g(1 - 0.68 D_p^{0.92} \rho_p^{0.5} \rho_f^{-0.2} D^{-0.54})$
Yang (1976)	$U_f - U_t \sqrt{\left(1 + \dfrac{f_p U_p^2}{2gD}\right)} \, e^{4.7}$

(1953) correlated the particle velocity with a number of the system parameters. A modified correlation has been found to work quite well when compared to some experimental systems. The convenience of this expression is notable. The third expression in Table 4-3 for the particle velocity is that of Yang. This expression works well for analysis, but it has an implicit format. Particle velocities using the various expressions are given in Example 4-4 for coal, limestone, and iron oxide.

Example 4-4 The particle velocity of a material in gas-solid transport is of much concern in analyzing the overall pressure losses in transporting the material. For 0.0254-m-diameter Schedule-40 pipe having air as a transport gas with a superficial gas velocity of 9.1 m/sec, determine the particle velocity using the terminal velocity analysis and Hinkle's correlation for comparison. Consider the following solid materials for analysis of the particle velocities:

Material		Sizes, μm		
Coal (Montana rosebud), Sp gr 1.2	10	100	10,000
Limestone, Sp gr 2.7	10	100	10,000
Iron oxide, Sp gr 5.7	10	100	10,000

In order to calculate the terminal velocities, the parameter K from Chap. Three is needed:

$$K = D_p \left(\frac{g\rho_g(\rho_p - \rho_f)}{\mu_f^2}\right)^{\frac{1}{3}}$$

$$\mu_f = 0.01 \text{ cP}$$

$$\rho_f = 1.2 \text{ kg/m}^3$$

$$\rho_{coal} = 1200 \text{ kg/m}^3$$

$$\rho_{limestone} = 2688 \text{ kg/m}^3$$

$$\rho_{Fe_2O_3} = 5705 \text{ kg/m}^3$$

Material	K value		
	10 μm	100 μm	10,000 μm
Coal	0.521	5.21	521
Limestone	0.671	6.81	681
Iron oxide	0.876	8.76	876

According to these values, for $K < 3.3$ one has Stokes' flow, and for $3.3 < K < 43.6$ one employs the intermediate flow range. For K values greater than 43.6, $C_D = 0.44$.

The terminal velocities for these ranges are

$$U_t \text{ (Stokes)} = \frac{gD_p^2 (\rho_p - \rho_f)}{18\mu_f}$$

$$U_t \text{ (intermediate)} = \frac{0.153g^{0.71}D_p^{1.14}(\rho_p - \rho_f)^{0.71}}{\rho_f^{0.29}\mu_f^{0.43}}$$

$$U_t \text{ (Newton)} = 1.75 \sqrt{\frac{gD_p(\rho_p - \rho_f)}{\rho_f}}$$

Therefore:

Material	U_t values, m/sec		
	10 μm	100 μm	10,000 μm
Coal	0.0065	0.438	17.3
Limestone	0.0146	0.777	25.9
Iron oxide	0.0311	1.33	37.8

Thus:

Material	$U_p = U_g - U_t$, m/sec		
	10 μm	100 μm	10,000 μm
Coal	9.1	8.66	—
Limestone	9.08	8.32	—
Iron oxide	9.07	7.77	—

For the modified Hinkle correlation,

$$U_p = U_g(1 - 0.68D_p^{0.92}\rho_p^{0.5}\rho_f^{-0.2}D_t^{-0.54})$$

	U_p, m/sec		
Material	10 μm	100 μm	10,000 μm
Coal	9.05	8.65	—
Limestone	9.02	8.43	—
Iron oxide	8.98	8.12	—

A comparison of the pressure drops for various frictional representations is given in Example 4-5. An additional effect of particle diameter is seen in Example 4-6.

Example 4-5 Using the same flow conditions as given in Example 4-3, determine the pressure drop for the frictional representations of Stermerding, Van Swaaij et al., Capes and Nakamura, and Yang, as given in Table 4-2. The first three expressions,

$$\text{Stermerding} \qquad f_s = 0.003$$

$$\text{Van Swaaij et al.} \qquad f_s = 0.080 \, U_p^{-1}$$

$$\text{Capes–Nakamura} \qquad f_s = 0.048 \, U_p^{-1.22}$$

can use the same analysis as given in Example 4-3. From these values, one has

$$\Delta P/L \text{ Stermerding} \quad = 5.09 \text{ kN/m}^2$$

$$\Delta P/L \text{ Van Swaaij et al.} = 6.44 \text{ kN/m}^2$$

$$\Delta P/L \text{ Capes–Nakamura} = 10.48 \text{ kN/m}^2$$

For the Yang frictional representation, one must solve a series of implicit expressions. A computer analysis is advisable if repeated calculations are to be made. A Newton–Raphson analysis may be performed on the Yang expressions in Tables 4-2 and 4-3 to obtain

$$U_p = U_f - U_t \sqrt{1 + \frac{f_p U_p^2}{2gD}} \, \epsilon^{4.7}$$

and

$$f_p = 0.0126 \frac{(1 - \epsilon)}{\epsilon^3} \left[(1 - \epsilon) \frac{Re_t}{Re_p} \right]^{-0.979}$$

This procedure yields

$$\Delta P/L \text{ Yang} = 5.62 \text{ kN/m}^2$$

Example 4-6 In many pressure-drop models for solid-gas flow the particle size of the solids is an important parameter; this average diameter may be determined in a number of ways. Using the Yang model for determining the pressure drop, vary the particle size to see now the pressure drop varies for the same case as Example 4-3. Use particle sizes 100, 50, 10, 5, and 1 μm.

In Yang's analysis the effect of diameter comes into play mainly through the terminal velocity, which is present in the particle velocity and friction factor representation. By analyzing the system for the various diameters of particles given above one finds:

$\Delta P/L$, N/m^2/m	D_p, μm
327	100
325	50
449	10
1,027	5
1,774	1

The smaller diameters increase the pressure losses seen in this pneumatic transport. As one may be able to appreciate, a difficult task is calculating precisely the particle size of the flowing material.

4-7   DENSE-PHASE AND MINIMUM-PRESSURE-DROP REGION IN VERTICAL FLOW

The preceding development has been given in rather general terms. Little was said as to the regime of flow in which the system operates. There is little difficulty in analyzing the dilute-phase regime by the proposed techniques, but some trouble occurs when the generalized approach is used for dense-phase gas-solid transport. A rather wide range exists among the predictions of the pressure drop for some models and actual compiled experimental data for the dense-phase region. This degree of departure is not seen in the dilute-phase regime. The basic equations for momentum transfer in the dense phase are correct, but undoubtedly something additional is happening to the solid-gas system that has not been accounted for. Capes and Nakamura (1973), in visually observing the flow of solids in a gas, note that a waviness of solid trajectories seen with the appearance that particles are floating near the wall. Recirculation of particles occurs moving down the wall and back into the core. These observations, among others, may be reason enough for the disagreement among models and experiments for the dense phase. Using the basic momentum analysis, the expression for the solid friction factor employed by Konno and Saito (1969) was analyzed to look at pressure-drop data in the vicinity of the minimum-pressure-drop region, as seen in Fig. 4-1. A comparison between the predicted pressure drop per unit length and experimental data is seen in Fig. 4-5. In carrying out these calculations the particle velocity as defined by expression 1 of Table 4-3 was employed. A little more than half of the data points considered could not be represented for this analysis for this minimum-pressure-drop region of flow. Other solid friction factor correlations yielded poorer results when compared to the experimental data, as seen in Table 4-4.

In performing the momentum transfer analysis, the particle velocity becomes incorporated into the final expressions through the solids friction factor. As mentioned before, determination of this solids velocity is difficult. For designing pneumatic flow

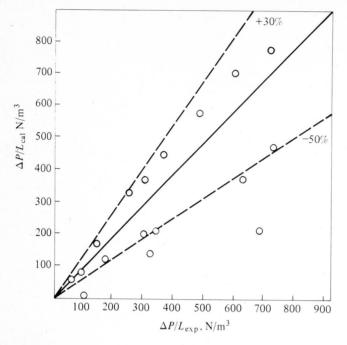

Figure 4-5 Calculated pressure drop per unit length versus experimental data employing the Konno–Saito model for the solid fraction factor.

systems it is suggested to use a correlation for the solid friction factor that does not depend on the solid particle velocity but on the gas velocity. Correlating the data from Table 4-4 by means of a multiple regression analysis gives the following two expressions for the solid friction factor:

$$f_s = a_1 U_g^{b_1} \tag{4-14}$$

$$f_s = a_2 U_g^{b_2} U_t^{b_3} \epsilon^{b_4} \tag{4-15}$$

where $a_1 = 0.021$ $a_2 = 0.049$
     $b_1 = -0.90$ $b_2 = -1.33$
              $b_3 = 0.139$
              $b_4 = 4.80$

The second expression incorporates two additional parameters for consideration—the terminal velocity and the voidage. Figure 4-6 shows the agreement of the experimental data with model values. For the minimum-pressure-drop range this technique is recommended for design of vertical transport systems.

Dixon (1976) has suggested using the minimum point in pressure drop with velocity for vertical pneumatic conveying in order to determine the solids friction factor and

Table 4-4 Experimental values of minimum $\Delta P/L$ and associated velocities

Process	Particle diameter D_p, μm	Particle density ρ_p, kg/m³	Tube diameter D, mm	Solids flow W_s, kg/sec	Gas density ρ_f, kg/m³	Total pressure P_t kN/m²	Velocity at minimum U_{min}, m/sec	Terminal velocity U_t, m/sec	Minimum pressure drop $\Delta P/L$, N/m²/m
Saroff et al. (1976), synthane PDU (coal)	100	1,200	7.62	0.0504	46.5	4,238	3.9	0.125	4,283
Zenz (1949), rape seeds	1,676	1,090	44.4	0.0295	1.2	101	9.1	0.77*	112
	1,676	1,090	44.4	0.0614	1.2	101	9.1	6.77*	200
	1,676	1,090	44.4	0.106	1.2	101	9.1	6.77*	359
	1,676	1,090	44.4	0.162	1.2	101	7.62	6.77*	363
	1,676	1,090	44.4	0.219	1.2	101	6.71	6.77†	785
glass	587	2,484	44.4	0.0204	1.2	101	8.53	4.36	94
	587	2,484	44.4	0.0910	1.2	101	7.92	4.36	358
	587	2,484	44.4	0.197	1.2	101	9.75	4.36	112
sand	930	2,484	44.4	0.010	1.2	101	8.23	7.71*	112
	930	2,484	44.4	0.0289	1.2	101	8.23	7.71*	179
	930	2,484	44.4	0.0727	1.2	101	9.1	7.71*	313
	930	2,484	44.4	0.241	1.2	101	9.75	7.77*	763
salt	167.6	2,099	44.4	0.017	1.2	101	4.88	0.927	61
	167.6	2,099	44.4	0.629	1.2	101	5.18	0.927	336
	167.6	2,099	44.4	0.108	1.2	101	6.4	0.927	719
Capes and Nakamura (1973), glass	2,896	2,853	76.2	0.0234	1.2	101	17.98	14.36*	79
	2,896	2,853	76.2	0.152	1.2	101	23.16	14.36*	159
Sandy, Daubert and Jones (1970) alumina	201	3,974	12.7	0.003	1.2	101	6.71	1.80	151
	201	3,974	12.7	0.009	1.2	101	8.53	1.80	317
Knowlton and Bachovchin (1975), lignite	363	1,260	73.7	0.428	5.63	483	14.0	0.99	263
	363	1,260	73.7	0.622	5.63	483	14.6	0.99	380
	363	1,260	73.7	0.866	5.63	485	15.2	0.99	501
	363	1,260	73.7	0.328	39.4	3,379	7.16	0.55	627
siderite	157	3,910	73.7	1.02	12.5	1,076	7.46	0.67	741
Konchesky, George, and Craig (1975), coal	50,900	1,506	137	0.731	1.2	101	32.9	43.8†	165
	50,900	1,506	137	1.26	1.2	101	39.0	43.8†	279
	50,900	1,506	137	2.67	1.2	101	41.1	43.8†	449
Rose and Barnacle (1957), glass, mustard seed, iron shot	1,996	1,142	31.7	0.031	1.2	101	9.45	7.56*	187
	1,996	1,142	31.7	0.0379	1.2	101	10.7	7.56*	178

*Small-particle velocities as defined by expression (1) in Table 4-3.
†Terminal velocity of the average particle size greater than gas velocity.

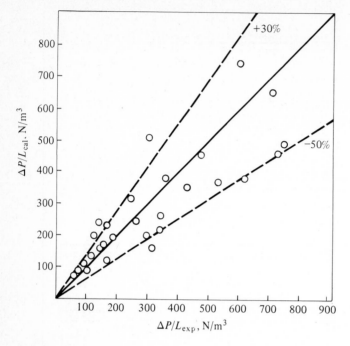

Figure 4-6 Calculated pressure drop per unit length versus experimental data for the solid friction factor $f_p = a_2 U_g^{b_2} U_t^{b_3} \epsilon^{b_4}$.

drag coefficient. Using Eq. 4-6 and setting the derivative $d(\Delta P/\Delta L/dU_p) = 0$, one obtains

$$U_{p(min)} = \left(\frac{2gD}{f_g}\right)^{\frac{1}{2}} \tag{4-16}$$

and

$$\frac{\Delta P}{\Delta L}_{(min)} = \frac{2W_s g}{(\pi/4)D^2 (f_s/2gD)^{1/2}} \tag{4-17}$$

From experimental determination of $(\Delta P/\Delta L)_{min}$ one can find f_s. At this minimum point one can also write

$$\frac{C_D}{C_{Dt}} \left[\frac{U_{G(min)} - U_{p(min)}}{U_t}\right]^2 = 2 \tag{4-18}$$

where C_{Dt} is the drag coefficient based on the terminal velocity of the particle.

Kerker (1977) has proposed a pressure-drop model that is dependent on a number of correlating factors. He employed his own data as well as that of several other investigators in his analysis. Analysis of Kerker's model with some dense-phase data in the literature has shown this model not to be applicable for these cases. Use of the model in the dilute-phase regime is possible, but the model has a disproportionate number of correlating parameters compared to other models for the dilute-phase regime.

Marcus and Vogel (1979) and coworkers have studied dense-phase transfer of solids achieving loading rates up to 500. By use of a conical throttle valve they were able to maintain a homogeneous flow having neither plugs nor dunes. They term this extremely dense flow extrusion flow.

Analyzing large loadings greater than 50 for horizontal transport has revealed an interesting application of flow of fluids through porous media to pneumatic transport. The concept is essentially that of Darcy's law relating the pressure drop through porous media to the fluid velocity. The factor of proportionality is that of the permeability. The data of Chari (1971) show that the permeability concept may be applied to high-loading pneumatic transport. The pressure drop per unit length in this case has two terms—that due to the gas flow through the moving porous media and the sliding friction term of the solids with the wall:

$$\frac{\Delta P}{L} = \frac{\mu}{K}(U_g - U_p) + \frac{\mu_s \rho_B U_p^2}{D} \tag{4-19}$$

where ρ_B = bulk density

$\quad K$ = permeability

The coefficient of sliding friction μ_s is assumed to be 0.36 from recent experiments. This permeability, K, can be related to the parameters of the system as the loading, tube diameter, and particle diameter:

$$K = 3.28 \times 10^{-14} \left(\frac{W_s}{W_g}\right)^{0.479} D^{-0.729} D_p^{0.248} \tag{4-20}$$

Figure 4-7 shows a comparison of pressure drops determined for the Chari data with the model given by Eq. 4-20. For high gas velocities where Darcy's law is not valid, the data of Albright et al. (1951) and Marcus and Vogel (1979) show the pressure drop per unit length to be proportional to $(U_g - U_p)^2$ with a factor of proportionality, the permeability, equal to $6.59 \times 10^{-4}(W_s/W_g)^{3.15} D^{0.357} D_p^{-0.84}$.

4-8 CHOKED-FLOW REGIME

Both choked flow and saltation are generally unstable conditions in the field of gas-solid transport. Most designers recommend avoiding these conditions, which means systems are designed to the right of the curves shown in Fig. 4-1. Both conditions involve instabilities, and the overall phenomenon is not clear-cut. Extrusion flow, which exists at high solid loadings and high pressure drops, is a stable flow of the moving-bed type that may exist beyond the choked-flow conditions.

An interesting coupling phenomenon occurs in a pneumatic transfer system involving a blower to move the gas-solid mixture. Doig (1975) and Leung, Wiles and

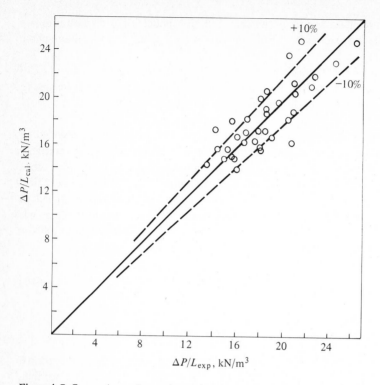

Figure 4-7 Comparison of experimental and calculated pressure drops determined using the permeability concept for Chari's data.

Nicklin (1971) point out this instability, which can be seen in Fig. 4-8. The blower pressure curve is seen superimposed on the two-phase pressure-drop curve. If the system is working at point *c,* which is the intersection of the pressure drop and performance curve of the blower, and some disturbance occurs such as a small blockage or surge, the system pressure can move to point *b* and then to point *a* if the disturbance is large enough. Point *a* is on the steep part of the pressure-drop curve and is rather unstable and easily driven to the choked-flow condition.

Some recent analyses have recommended the behavior of the gas-solid system at or near the choked-flow condition. Leung, Wiles, and Nicklin (1971) have proposed a correlation for choking conditions that assumes the voidage at the choked condition is 0.97. Capes and Nakamura, in their experimental analysis of choked flow, found the voidage at choked conditions to be closer to 0.99 than the value Leung, Wiles, and Nicklin suggest. Leung and coworkers' analysis then gives the volumetric solid flow rate per unit cross section (superficial solid velocity) at the choked condition as

$$U_s' = \frac{U_g - 0.97 U_t}{32.3} \tag{4-21}$$

with U_g as the superficial fluid velocity.

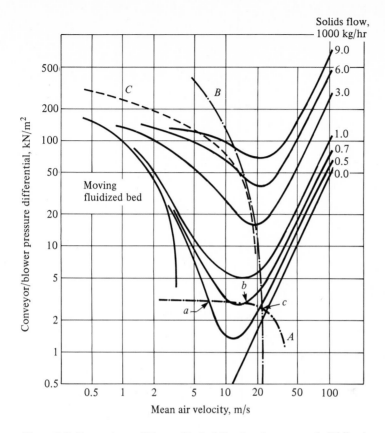

Figure 4-8 Comparison of blower (dashed lines) and conveyor (solid lines) pressure-drop relationships (Doig, 1975). Curve A, radial bladed fan; curve B, isothermal expansion at constant mass flow; curve C, positive displacement blower; see text for explanation of points a, b, and c. (Reproduced by permission of The South African Institution of Mechanical Engineering.)

Knowlton and Bachovchin (1975) have studied gas-solid pressure-drop velocity at the choking conditions. Their study was performed on coal at high pressures of 482 to $3379 k$ N/m^2. Multiple regression analysis was performed on the variables to determine a correlation for the choking velocity:

$$\frac{V_{choke}}{\sqrt{gD_p}} = 9.07 \left(\frac{\rho_p}{\rho_f}\right)^{0.347} \left(\frac{W_s D_p}{\mu_f}\right)^{0.214} \left(\frac{D_p}{D}\right)^{0.246} \tag{4-22}$$

This correlation was in agreement with the data to within 10%.

Yang (1976) has correlated the friction factor for the solids using his equation for the particle velocity,

$$U_p = U_f - U_t \sqrt{\left(1 + \frac{f_p U_p^2}{2Dg}\right) \epsilon^{4.7}} \tag{4-23}$$

and the value of the particle velocity at the choked condition,

$$U_p = U_f - U_t \tag{4-24}$$

noting that this friction factor is approximately equal to 0.01 at this condition. Thus, this analysis gives a friction factor representation as

$$f_{p\text{(choked)}} = 0.01 = \frac{2D_g(\epsilon^{-4.7} - 1)}{(U_f - U_t)^2} \tag{4-25}$$

The amount of solids flowing at this choked condition can then be expressed as

$$\frac{W}{A}_{\text{(choked)}} = (U_{f\text{(choked)}} - U_t)\rho_p (1 - \epsilon_{\text{choked}}) \tag{4-26}$$

It is noteworthy that Yang's analysis includes some more recent data of Yousfi and Gau (1974).

4-9 SALTATION

In horizontal flow of solids and gas, saltation occurs when the carrier gas velocity is small enough to permit settling of the solids particles within the transport line. Wen and Simons (1959) have studied the various stages of saltation until a completely plugged pipe results. Figure 4-9 shows the results of the various stages of saltation as

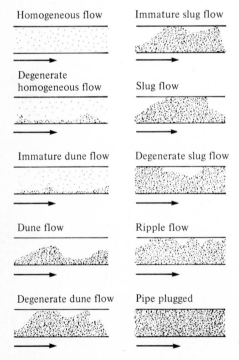

Figure 4-9 Observations of the various flow patterns (Wen and Simons, 1959). Reproduced by permission of *AIChE Journal*.

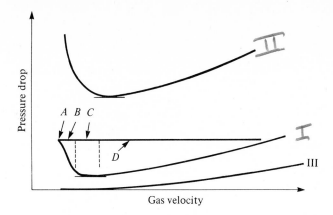

Figure 4-10 Description of observed patterns of particle behavior in a horizontal pipe (Owens, 1969). Curve I, light solid loading; curve II, heavy solid loading; curve III, particle-free gas; *A*, choking; *B*, deposition, dune formation, saltation; *C*, higher concentration in lower part of pipe; *D*, uniform suspension. (Reproduced by permission of *J. of Fluid Mechanics,* Pergamon Press.)

visually observed by Wen and Simons. In these types of flows, the larger particles segregate toward the bottom of the pipe. Some descriptions of the complex flow situations that exist at the various stages of saltation have been attempted. In addition to these solid deposits, Owens (1969) shows the relation between the pressure drop as compared to the conditions occurring within the pipe (see Fig. 4-10).

Adam (1957) has studied the particle trajectories of solids in pneumatic transfer lines by use of high-speed motion pictures. Adam found that the particles made two types of collisions with the wall. The first type is an elastic bounce of a rapidly spinning particle. The second type of collision is characterized by a deformation of the particle upon collision or an indentation in the pipe wall due to the impact. As a result of these collisions, irregular bouncing along the pipe wall is expected.

The analyses given previously by application of Yang's equations for horizontal flow will hold only for the cases of dilute homogeneous flow of solids and gas. Care must be taken not to apply the dilute-phase equations when saltation of varying degrees exists in the flow system.

Owens (1969) has suggested that inspection of the dimensionless ratio of fluid momentum to particle gravitation force (a pseudo-Froude number),

$$\frac{\rho_f \bar{U}^2 f_g}{2\rho_p g D_p}$$

may help determine when saltation is likely to occur. Owens suggests that saltation is of importance when this parameter is between 1 and 10^{-2}. For values less than 10^{-2} large depositions are anticipated.

A useful design parameter for horizontal transfer of solids by a gas stream is the minimum superficial velocity required to convey the particles. Zenz (1964) has done

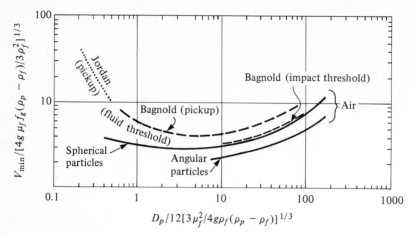

Figure 4-11 Minimum transport velocity versus particle diameter for horizontal flow (Zenz, 1949). Reproduced by permission from American Chemical Society.

an analysis of the minimum transport velocity for single particles and compared this with experimental data on multiparticle systems studied by Bagnold (1941) and Jordan (1954). Figure 4-11 shows the plot of this minimum velocity versus the particle size.

Thomas (1961, 1962) has developed a correlation base on the friction velocity principle. The technique for determining the minimum gas velocity for saltation involves various implicit expressions. The gas velocity is given as

$$\frac{U_s}{U_s^*} = 5 \log Re - 3.90 \tag{4-27}$$

where $Re = \dfrac{\rho_f D U_s}{\mu_f}$

$$U_s^* = \left(\frac{\Delta P D}{4 L \rho_\mathrm{m}}\right)^{\frac{1}{2}}$$

ρ_m = mixture density

The value of the friction velocity at saltation for a suspension as related to friction velocity at saltation of a single particle is

$$\frac{U_s^*}{U_{s0}^*} = 1 + 2.8 \left(\frac{U_t}{U_{s0}^*}\right)^{\frac{1}{2}} (1 - \epsilon_s)^{\frac{1}{2}} \tag{4-28}$$

where U_t = terminal velocity of a single particle

ϵ_s = voidage at saltation, assumed to be when $U_s/U_g = 0.5$

The term U_{s0}^*, the frictional velocity at saltation for a single particle, can then be found from one of two equations, depending on the size of the particle. For $D_p \geqslant 5\mu f/\rho f U^* = \lambda$.

$$\frac{U_t}{U_{s0}^*} = 4.90 \left(\frac{D_p U_{s0}^* \rho_f}{\mu_f}\right) \left(\frac{DU_{s0}^* \rho_f}{\mu_f}\right)^{-0.6} \left(\frac{\rho_p - \rho_f}{\rho_f}\right)^{0.23} \tag{4-29}$$

and for $D_p < \lambda$,

$$\frac{U_t}{U_{s0}^*} = 0.01 \left(\frac{D_p U_{s0}^* \rho_f}{\mu_f}\right)^{2.71} \tag{4-30}$$

4-10 DESIGN RECOMMENDATIONS

For dilute-phase vertical transport the Institute of Gas Technology (1978) team on analysis of various investigators' data on coal and coal-related materials recommends the Konno–Saito model (Table 4-2) for prediction of the pressure drop. The author has found generally that both the Konno–Saito and Yang models are about equally valid in analyzing dilute-phase systems. The Konno–Saito model as analyzed by the IGT team includes the acceleration effects and defines the solid velocity by expression 1 of Table 4-3. In the IGT analysis the acceleration effects as defined by Yang were also included in the calculations. The overall deviation between the measured and predicted values of the pressure drop for the IGT, Konno–Saito analysis was 30%, while it was 45% for the Yang model.

The IGT team also analyzed horizontal dilute-phase gas-solid transfer. They found the Yang model and a modified Hinkle analysis to work about equally well for prediction of pressure drop on coal and coal-related materials. The modified Hinkle analysis involves modification of the particle velocity developed by Hinkle. The particle velocity that gave minimum deviations between the experimental and predicted pressure drop is

$$U_p = U_g \left(1 - 0.044 D_p^{0.3} \rho_p^{0.5}\right) \tag{4-31}$$

The pressure drop from Hinkle's model is given as

$$\frac{\Delta P}{L} = \frac{U_g^2 \rho_f}{2L} + \frac{W_s U_p}{LA} + \frac{2f_g \rho_f U_g^2}{D} + \frac{2f_s U_p W_s}{DA} \tag{4-32}$$

with

$$f_s = \frac{3}{8} \frac{\rho_f}{\rho_p} C_D \frac{D}{D_p} \left(\frac{U_g - U_p}{U_p}\right)^2 \tag{4-33}$$

For dense-phase design of pressure drop, recent work has treated data having loadings from 10 to 50 from a number of studies to predict the solid friction factor. The results of analyzing experimental and calculated pressure drops show a ±20% variation. The solid friction factor from this analysis is

$$f_p = \frac{0.146 D^{1.07}}{U_g^{0.643} D_p^{0.259} \rho_p^{0.909}} \tag{4-34}$$

A standard error estimate of 0.291 and a correlation coefficient of 0.901 is obtained by use of this equation for dense-phase flow.

A current analysis by Leung has subdivided the dense-phase regime into two branches: one with slugging and the other with nonslugging followed by slugging. The parameters of concern are the loadings and the gas and solid velocities.

Jones and Leung (1978) have done an extensive analysis on saltation velocities in pneumatic conveying. They have concluded that Thomas's (1961) (1962) correlation, which is based on a fundamental saltation mechanism, is best. The use of Thomas's correlation is fairly complicated, because it involves some implicit expressions.

4-11 CONCENTRATIONS OF PARTICLES IN A TUBE

Some confusion may initially arise when one views the data on concentration profiles in a tube. At first it may appear that there are contradictory results, but the conditions under which the experiments are performed must be looked at closely. Apparent contradictions may exist in many aspects of gas-solid flow if a very narrow range of parameters is viewed by separate investigators.

In a 0.13-m-diameter pipe with air flowing at 40 m/sec, Soo et al. (1964) found that 50-μm-diameter particles concentrate in the wall region. Arandel, Bibb, and Boothroyd (1970) found similar behavior for 15-μm particles in a 0.076-m-diameter duct. Looking at acceleration development lengths, Zenker (1972) studied various sizes of particles in a 0.041-m-diameter pipe. For 42-μm-diameter particles, Zenker found a high concentration of particles at the wall. When studying 238-μm-diameter particles, the concentration of particles shifted to a maximum in the center. Particles of 133 μm diameter showed a concentration profile somewhere in between the other profiles.

Kramer (1970) has studied 62- and 200-μm-diameter particles at Reynolds number 25,500 in 0.0127-m-diameter tubes. Here the results were opposite to those of Zenker, showing higher concentrations in the center for 62-μm-diameter particles and a more uniform concentration distribution for the 200-μm-diameter particles. Kramer's

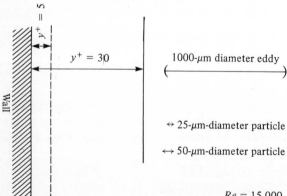

Figure 4-12 Comparison of particle and eddy sizes.

results may well be interpreted in view of the other investigators' with an emphasis on the small tube diameter and lower gas flow Reynolds number.

Actually, the behavior of solid particles in the flow field is quite complex, depending on the structure of turbulence. Of interest in this respect is the ejection sequence viewed by Corino and Brodkey (1969) in their study of single-phase flow. The interaction of fluid eddies with particles is more likely as the size of the particles is decreased. This injection sequence could well keep the large particles at the center of the pipe and concentrate the smaller particles that follow the turbulence more closely in the wall region. Figure 4-12 shows the relative sizes of particles and eddies.

4-12 PARTICLE ADDITION TO A GAS STREAM FOR DRAG REDUCTION

A number of investigators have found that the addition of small-diameter particles to a flowing gas stream results in drag reduction (Boyce and Blick, 1969; Peters and Klinzing, 1972; Pfeffer and Rossetti, 1972; Soo and Trezek, 1966). Economically, this means that it costs less in energy expenditure to move the two-phase flow system than the single-phase system. Saying that the addition of mass to a flow should result in more economical energy usage goes against one's intuition, but Fig. 4-13 shows experimental

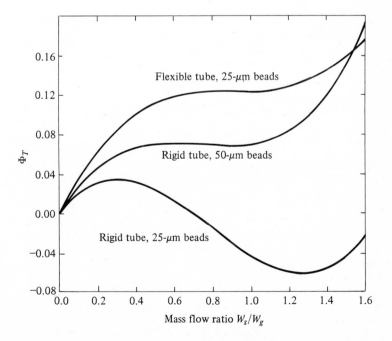

Figure 4-13 Fractional change in the shear stresses resulting from the presence of solid particles as a function of mass flow ratio (Peters and Klinzing, 1972).

verification of this phenomenon. The parameter Φ_T represents the change in shear stress due to the presence of particles. When Φ_T is negative, a drag reduction is seen. The particle size is a crucial parameter for this phenomenon; the diameters of about 20 to 75 μm are the average sizes of the particles over which the phenomenon of drag reduction is seen by all investigators. The solid to gas loading must be in the regime of about 0.5 to 4, and the gas Reynolds numbers for the phenomenon are in the turbulent region from about 15,000 to 40,000.

Some interesting LDV measurements have been made on such drag-reducing systems to yield a further understanding of the reasons for the phenomenon. In the past, the interaction between the particle and the fluid turbulence or turbulent eddies has been suggested as a possible reason for the phenomenon. Carlson (1973) and Kolansky (1976) have recorded LDV data on gas-solid systems exhibiting drag-reducing properties. Both have found the turbulence intensities for the solid to be greater than the gas in the wall region. Carlson attributes this character to the polydispersity of the solids employed in the study. Kolansky has attempted to describe this behavior in light of newer experiments and theories. These theories arise from the works of Kline et al. (1967), Clark and Markland (1970), and Corino and Brodkey (1969). Particles that are excited by bursts from the sublayer, like the eddies that accelerate them, terminate their trajectories in the buffer zone. The slip between the two phases is due to the local interaction of the particles with the eddies. The effect of the particles on the gas is to thicken the buffer layer by drawing energy from the entire flow by the Stokes dissipation that results from the particle slip and to increase the turbulence intensity.

Some very interesting drag-reduction studies have been performed by Marcus (1979), coupling gas-solid flow with high-frequency oscillations superimposed on the flow system. A length dependency was found on the phenomenon of drag reduction, pointing to possible analysis by finite amplitude or acoustical theory to the two-phase flow system.

PROBLEMS

4-1 Using the Konno–Saito expression for frictional representation for gas-solid flow, determine the pressure drop per unit length for the vertical coal flow of 0.0126 kg/sec of 100-μm-diameter particles through a 0.0095-m Sh 40 pipe at 689.5 kN/m^2 pressure. Consider the gas (nitrogen) velocity to be 2.4 m/sec. What is the pressure drop per unit length for 6.1 and 11 m/sec velocities? In an analysis of these cases state their advantages and disadvantages.

4-2 For the conditions in Prob. 4-1, determine the acceleration lengths considering the two transport gas velocities. Assume the particle velocity is introduced into the system at 0.305 m/sec.

4-3 Develop a generalized numerical procedure for solution of the implicit Yang expressions in determining the pressure losses in vertical gas-solid transfer.

4-4 For the transport of coal at 0.063 kg/sec in a 0.0127-m-diameter line, the transport gas velocity is varied from 2.1 to 9.1 m/sec. Using the Capes–Nakamura representation for the solid frictional contribution, construct a plot of the pressure drop per unit length using varying gas velocities. Consider the gas to be nitrogen at 207 kN/m^2 pressure and 20°C. The average particle diameter of the coal is 150 μm, and the flow is vertical. How would this behavior look for horizontal flow?

4-5 For a dense-phase flow of coal at 0.126 kg/sec in a 0.0095-m Sh 40 pipe of 100-μm-diameter particles with a nitrogen transport velocity of 5.49 m/sec, determine the pressure drop per unit length for various frictional representations. The flow is vertical and the gas pressure is 689.5 kN/m^2 at a temperature of 21°C.

4-6 Consider the flow systems listed in the table. Determine whether saltation is of importance in the horizontal flow of gas-solids. The transport gas is nitrogen in a 0.0245 m diameter pipe.

material	Particle diameter, μm	Mean velocity, m/sec	Gas pressure, kN/m^2
Glass	150	20	101
Lead	10	20	101
Coal	250	15	345
Urania	20	17	101
Iron oxide	50	15	310
Coal	100	14	689.5

4-7 Consider the flow of glass beads of an average particle diameter of 26 μm flowing in a 0.0254-m-diameter pipe at a solids/gas ratio of 1.2. Determine the pressure drop per unit length for this system with and without drag reduction present. The ratio of the friction factor based on a mixture density to the friction factor based on gas density is given as

$$\frac{f_m}{f_g} = \frac{1 + \Phi_T}{1 + W_s/W_g}$$

The Reynolds number of the gas is 20,000.

4-8 For a coal flow of 0.025 kg/sec and average particle diameter of 200 μm in a 0.0254-m-ID pipe at a carrier gas pressure of 207 kN/m^2, determine the gas velocity that will give the choked-flow condition in vertical transport.

REFERENCES

Adam, O.: *Chem. Eng. Tech.* **29**:151 (1957).

Albright, C. W., J. H. Holden, H. P. Simons, and L. D. Schmidt: *IEC* **43** 1837 (1951).

Arundel, P. A., S. D. Bibb, and R. G. Boothroyd: Powder Technology **4**:302 (1970).

Bagnold, R. A.: *The Physics of Blown Sand and Desert Dunes,* Methuen, London, 1941.

Boothroyd, R. D.: *Trans. Inst. Chem. Eng.* **44**:T306 (1966).

Boyce, M. P., and E. F. Blick: "Fluid Flow Phenomena in Dusty Air," ASME Paper no. 69-WA/FE-24 (1969).

Capes, C. E., and K. Nakamura: *Can. J. Chem. Eng.* **51**:31 (1973).

Carlson, C. R.: "Turbulent Gas-Solids Flow Measurements Utilizing a Laser Doppler Velocimeter," Ph.D. Thesis, Rutgers University, New Brunswick, NJ, 1973.

Chari, S. S.: *AIChE Symp.* Series 116 **67**:77 (1971).

Clark, J. A., and E. Markland: *Aeronaut. J.* **74**:243 (1970).

Corino, E. R., and R. S. Brodkey: *J. Fluid Mech.* **37**:1 (1969).

Dixon, G.: *Int. J. Multiphase Flow* **2**:465 (1976).

Doig, I. D.: *S. African Mech. Eng.* **25**:394 (1975).

Hinkle, B. L.: "Acceleration of Particles and Pressure Drops Encountered in Horizontal Pneumatic Conveying," Ph.D. Thesis, Georgia Institute of Technology, Atlanta, GA, 1953.

Hjulstrom, F.: *Bull. Geol. Inst. Upsala* **23** (1936)

Institute of Gas Technology, Dept. of Energy Contract, FE 2286-32 (Oct. 1978).

Jones, P. J., and L. S. Leung: *IEC Process Des. Dev.* **17**:571 (1978).

Jordan, D. W.: *Brit. J. Appl. Phys.* **5** (Suppl. 3): S194–S198 (1954).

Kerker, L.: Thesis, "Drukverlust und Partikelgeschwindigkeit Gas Festoffstromung", University Karlsruhe, 1977.

Kline, S. J., W. C. Reynolds, F. A. Schraub, and P. W. Rundstadler: *J. Fluid Mech.* **30**:741 (1967).

Knowlton, T. M. and D. M. Bachovchin: International Conference on Fluidization, Pacific Grove, CA, June 15–20 (1975).

Kolansky, M. S.: Ph.D. Thesis, *Studies in Fluid Mechanics,* City University, New York, 1976.

Konchesky, J. L., T. J. George, and J. G. Craig: *Trans. ASME J. Eng. Ind.* Ser. B 97 (1) 94 Feb. 1975).

Konno, H., and S. Saito: *J. Chem. Eng. Japan* 2:211 (1969).

Kramer, T. J.: Ph.D. Thesis, "A Study of the Mean Flow Characteristics of Gas-Solid Suspensions Flowing in Vertical Tubes," University of Washington, Seattle, 1970.

Leung, L. S., and R. J. Wiles: *IEC Process Des. Dev.* 15:552 (1976).

—— and D. J. Nicklin: *IEC Process Des. Dev.* 10:188 (1971).

Marcus, R. D. and R. Vogel: *Proceedings Powder and Bulk Solids Conference,* 92, May 1979 (published by Industry and Scientific Conference Management, Inc., Chicago IL).

Owens, P. R.: *J. Fluid Mech.* 39:407 (1969).

Peters, L. K., and G. E. Klinzing: *Can. J. Chem. Eng.* 50:441 (1972).

Pfeffer, R., and S. J. Rossetti: *AIChE J.* 18:31 (1972).

Reddy, K. V. S., and D. C. T. Pei: *I.E.C. Fund.* 8:490 (1969).

Rose, H. E., and H. E. Barnacle: *The Engineer* 898 (June 14, 1957).

Sandy, C. W., T. F. Daubert, and J. H. Jones: *Chem. Eng. Prog.* 66(105):133 (1970).

Saroff, L., F. H. Gromicko, G. E. Johnson, J. P. Strakey, and W. P. Haynes: *69th Annual Meeting AIChE,* Chicago, Dec. 1976.

Shimizu, A., R. Echigo, and S. Hasegawa: *Int. J. Multiphase Flow* 4:53 (1978).

Soo, S. L., and G. J. Trezek: *Ind. Eng. Chem.* 5:388 (1966).

——, R. C. Dimick, and G. F. Hohnstreiter: *IEC Fund.* 3:98 (1964).

Stemerding, S.: *Chem. Eng. Sci.* 17:599 (1962).

Thomas, D. G.: *AIChE J.* 7:432 (1961); 8:373 (1962).

Van Swaaij, W. P. M., C. Buurman, and J. W. van Breugel: *Chem. Eng. Sci.* 25:1818 (1970).

Wen, C. Y., and H. P. Simons: *AIChE J.* 5:263 (1959).

Yang, W. C.: *International Powder and Bulk Solids Handling and Processing Conference Exposition,* Chicago, May 11–14, 1976. (Published by Industry and Scientific Conference Management, Inc.)

——: *AIChE J.* 24:548 (1978).

Yousfi, Y., and G. Gau: *Chem. Eng. Sci.* 29:1939 (1974).

Zenker, P., *Staub-Reinhalt, Luft.* 32:1 (1972)

Zenz, F. A.: *Ind. Eng. Chem.* 41:2801 (1949).

——: *IEC Fund.* 3:65 (1964).

FIVE

CURVES AND BENDS

5-1 INTRODUCTION

The topic of curves and bends in piping systems is often forgotten in pneumatic gas-solid transport, but only rarely can one find a straight pipe from one processing unit to another. Space limitations usually make the use of bends and curves in the piping essential.

The easiest method of calculating energy losses in bends and curves is to obtain a factor by which the diameter of the pipe may be multiplied in order to find an equivalent length of straight pipe for analysis. Indeed, this procedure exists for a single-phase flow and has been extrapolated for two-phase flow systems. This sort of analysis ignores the details of the behavior of the fluid and particles in bends and curves. Secondary flows and centrifugal forces on the particles are lost from consideration in this black box approach, but these forces and erosion have fortunately been explored to give us a more physical picture of this phenomenon of flow through bends. These studies can guide design and often suggest modifications to reduce energy losses, increase operability, and reduce erosion.

5-2 SINGLE-PHASE FLOW

Any engineering or fluid-flow handbook contains tables of equivalent lengths of straight pipe for piping designs. These are based on the type of bend, its radius of curvature, and, often, the conditions of the inner surfaces of the pipe. Table 5-1 shows a suggested set of factors for bend and curve designs (Perry, 1950). More recently, a tremendous amount of work has been carried out by Ito (1959) to determine pressure losses in pipe bends. A typical pressure-loss curve as seen by Ito when

Table 5-1   Friction loss of screwed fittings, valves, etc. (Perry, 1950)*

Factor	Equivalent length in pipe diameter L_e
45° elbow	15
90° elbow, standard radius	32
90° elbow, medium radius	26
90° elbow, long sweep	20
90° square elbow	60
180° close return bend	75
180° medium radius return bend	50
Tee (used as elbow, entering branch)	90
Coupling	Negligible
Union	Negligible
Gate valve, open	70
Globe valve, open	300
Angle valve, open	170
Water meter, disk	400
Water meter, piston	600
Water meter, impulse wheel	300

*From *Chemical Engineers' Handbook 3/edition,* copyright (1950, Perry). Used with permission of McGraw-Hill Book Company.

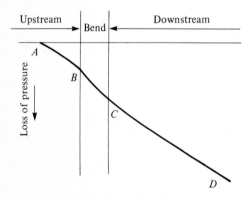

AB = Pressure loss before bend
BC = Pressure loss in bend
CD = Pressure loss after bend

Figure 5-1 Diagram of a typical hydraulic gradient in a bend. [Reproduced with permission from Ito, H., *Trans. ASME J. Basic Engr.* **82D**:131 (1960)]

a fluid traverses a bend is shown in Fig. 5-1. Ito has correlated his own data as well as that of several investigators to suggest the friction factor for bends in terms of the Dean number, $Re(r_0/R_B)^{1/2}$. In this expression, R_B is the radius of curvature of the bend and r_0 is the radius of the pipe. For fully developed turbulent flow in pipes, Ito suggests a friction factor and pressure-drop expression,

$$4f = \left\{ 0.029 + 0.304 \left[Re \left(\frac{r_0}{R_B} \right)^2 \right]^{-0.25} \right\} \left(\frac{R_B}{r_0} \right)^{-1/2} \tag{5-1}$$

or

$$\Delta P = \frac{\left\{ 0.029 + 0.304 [Re(r_0/R_B)^2]^{-0.25} \right\} (L\rho\bar{U}^2/2D)}{(R_B/r_0)^{1/2}} \tag{5-2}$$

where L is the axial length of the curve. These expressions give good agreement with experimental results from $300 > Re(r_0/R_B)^2 > 0.034$. Below $Re(r_0/R_B)^2 = 0.034$ the friction factor coincides with that of a straight pipe. For variable angle bends Ito (1960) further suggests

$$\Delta P = 0.00873\alpha(4f)\theta_2 \frac{R_B}{r_0} \rho \frac{\bar{U}^2}{2} \tag{5-3}$$

for $Re(r_0/R_B)^2 < 91$, where f is the friction factor from Eq. 5-1, θ_2 is the angle of inclination, and α is a numerical coefficient. For $Re(r_0/R_B)^2 > 91$,

$$\Delta P = 0.00241\alpha\theta_2 Re^{-0.17} \left(\frac{R_B}{r_0} \right)^{0.84} \left(\frac{\rho\bar{U}^2}{2} \right) \tag{5-4}$$

The value of α varies with the angle θ_2. When $\theta_2 = 45°$,

$$\alpha = 1 + 14.2 \left(\frac{R_B}{r_0} \right)^{-1.47} \tag{5-5}$$

when $\theta_2 = 90°$,

$$a = 0.95 + 17.2 \left(\frac{R_B}{r_0} \right)^{-1.96} \qquad \frac{R_B}{r_0} < 19.7 \tag{5-6}$$

$$a = 1 \qquad \frac{R_B}{r_0} > 19.7 \tag{5-7}$$

and when $\theta_2 = 180°$,

$$a = 1 + 116 \left(\frac{R_B}{r_0} \right)^{-4.52} \tag{5-8}$$

Examples 5-1 and 5-2 show some typical pressure drops through bends for a one-phase system.

Example 5-1 Consider the flow of nitrogen through a 0.0508-m-diameter line at 1013 kN/m^3. Determine the pressure drop around a 90° bend on a 0.61-m-radius and on a 0.305-m radius for a gas velocity of 6.1 m/sec.

For a 0.608-m radius of curvature, $L = 0.957$ m, and for a 0.304-m radius, $L = 0.479$ m. The Reynolds number is

$$Re = \frac{0.0508 \text{ m} \times 6.1 \text{ m/sec} \times 12 \text{ kg/m}^3}{10^{-8} \text{ kg/(m·sec)}} = 3.72 \times 10^5$$

The grouping $Re(r_0/R_B)^2$ can now be determined:

$$Re\left(\frac{r_0}{R_B}\right)^2_{(0.61\text{ m})} = 3.72 \times 10^5 \left(\frac{0.0254}{0.957}\right)^2 = 262$$

$$Re\left(\frac{r_0}{R_B}\right)_{(0.305\text{ m})} = 3.72 \times 10^5 \left(\frac{0.0254}{0.479}\right)^2 = 1046$$

The pressure drop is given by Eq. 5-2:

$$\Delta P_{(0.61\text{ m})} = 359 \text{ N/m}^2$$

$$\Delta P_{(0.305\text{ m})} = 201 \text{ N/m}^2$$

Example 5-2 For the same conditions as outlined in Example 5-1 use Table 5-1 to solve the problem considering a medium- and long-sweep radius.

In general, the pressure drop can be given as

$$\Delta P = \frac{2f L \bar{U}^2 \rho}{D}$$

For this case of turbulent flow the friction factor is assumed to be 0.004. For a long-sweep radius,

$$L_e = 20D = 20(0.0508 \text{ m}) = 1.02 \text{ m}$$

For a medium-sweep radius,

$$L_e = 26D = 26(0.0508 \text{ m}) = 1.32 \text{ m}$$

$$\Delta P_{\text{long}} = \frac{2(0.004)\,(1.02)\,(37.2)\,(12)}{(0.0508)} = 71 \text{ N/m}^2$$

$$\Delta P_{\text{medium}} = \frac{2(0.004)\,(1.32)\,(37.2)\,(12)}{(0.0508)} = 93 \text{ N/m}^2$$

These values seem to underestimate the values suggested by Ito as shown in Example 5-1. If a design were based on these approximations, overdesign would occur.

5-3 TWO-PHASE FLOW

A physical picture of what happens to the solids in a gas-solid flow system as they approach and leave a bend is of interest. Assuming the overall flow to be at steady state, as the gas-solid flow approaches a bend the solid and gas flows decelerate. The solid and gas velocities will go to zero if they collide directly with the perpendicular surface of a right-angle turn. Undoubtedly, deflections of the solid particles take place that do not allow the solid velocity go to zero. If the bend is less than 90°, then the degree of deceleration decreases as the angle decreases. Mason and Smith (1972) have provided some interesting photographs of gas-solid flow in bends that show the

Table 5-2 Flow, particle, and tube properties studied in investigations on bends

Investigator	Particle type	Particle size μm	Particle density, kg/m^3	Loading	Tube diameter, m
Schuchart (1968)	Glass	1,500–3,000	2,610	0–20	0.034
	Plastic	2,180	1,140		
Sproson, Gray, and Haynes (1973)	Coal	12,800	1,143	2.8–6.7	0.126
Mason and Smith (1973)	Alumina	15	3,990	0.4–4.8	0.051
		40			0.076
		70			
Morikawa et al. (1978)	Polyethylene	1,100	923	0–8	0.04

concentration of the particles on the outer regions of the bend due to the centrifugal force imparted to the particles by the flow around the bend. If the particles are small, their concentration on the outer radius is less than that of larger particles, since the centrifugal force is directly dependent on the mass of the particles.

After the particles have been decelerated by the bend, they must be accelerated to their previous steady-state value (before the bend). The energy loss in flow through a bend is made up of the contributions mentioned above. Little experimental data existed for pressure losses in gas-solid flow in bends and curves until several studies were finally performed; Table 5-2 is a compilation of the parameters studied. Schuchart's (1968) work on bends and curves is quite a complete study. These findings are seeing more acceptance with the increase in the data in the literature.

Sproson, Gray, and Haynes' (1973) study on large coal particles (12 mm in diameter) through bends agrees quite well with the predictions of Schuchart.

A study of pressure drop in right angles and elliptical bends by Morikawa et al. (1978) suggests a form for the pressure drop similar to Schuchart's. This model does not predict other investigators' data with much accuracy, although Mason and Smith's (1973) data on fine particles are more closely predicted than by other correlations.

The Mason and Smith (1973) work on the flow of fine powders through wide bends tested particles of 15, 40, and 70 μm in diameter for bends having a diameter ratio D_B/D of 20. Their data were overestimated by all the existing correlations for pressure drops in bends. A particle size effect is undoubtedly in order as a modification on such correlations of Schuchart or Morikawa.

Ikemori and Munakata (1973) experimentally found an expression for the bend frictional loss in terms of the particle velocity/gas velocity ratio as well as other factors. Use of this model appears to be more cumbersome than Schuchart's model.

In treating pressure losses in bends, Rose and Duckworth (1969) have suggested an empirical technique that considers only the acceleration portion of the pressure loss. Other analyses such as that of Yang may also be employed to calculate this acceleration pressure loss.

5-4 A DESIGN SUGGESTION

A suggested design is to use the single-phase results as a basis for the two-phase design. The pressure drop for a straight section of pipe having a gas-solid flow can be easily determined by a number of equations suggested by Eq. 4-6 to 4-9. Utilizing this information, one can now rely on Table 5-1 for equivalent lengths of straight pipe for bends to find the total equivalent length of a piping system. This length then can be used as the length term in the basic two-phase pressure-drop expressions. This technique has been used by a number of industrial designers with apparent success. Example 5-3 shows an application of this principle.

Example 5-3 Consider the conditions of Example 4-3 for the transport of coal. The pressure drop per unit length is given as $113.7 \text{ N/m}^2/\text{m}$. Assume that a system has 15.24 m of vertical length and ten $90°$ elbows of standard radius and gate valve in the open position. Estimate the total pressure drop for the system using the design suggestion procedures.
    From Table 5-1,

$$90° \text{ elbow of standard length} = 32D = L_{e1}$$

$$\text{Gate valve open} = 70D = L_{e2}$$

Therefore,

$$\Delta P_{\text{total}} = (113.7 \text{ N/m}^2/\text{m}) [15.24 \text{ m} + (10)(32)(0.0254 \text{ m}) + (4)(70)(0.0254 \text{ m})]$$

$$= (113.7 \text{ N/m}^2/\text{m})(30.5 \text{ m})$$

$$= 3.47 \text{ kN/m}^2$$

5-5 EXPERIMENTAL FINDINGS

Schuchart has performed a number of experiments on pressure losses in bends and curves for gas-solid flow. His experiments involve the use of glass and plastic particles between 1 and 2 mm in diameter with volumetric concentrations up to 5%. Schuchart gives the following expression for curves of varying degrees of curvature:

$$\frac{\Delta P_{\text{bend}}}{\Delta P_{\text{straight}}} = 210 \left(\frac{2R_B}{D} \right)^{-1.15} \tag{5-9}$$

where $\Delta P_{\text{straight}} = $ pressure drop of an equivalent length of straight pipe

$$R_B = \text{radius of curvature of the bend}$$

$$D = \text{diameter of the pipe}$$

For a $90°$ sharp elbow he suggested

$$\Delta P_{\text{bend}} = 12\left(\frac{\rho_p}{\rho_f}\right)^{0.7}\left(\frac{\rho_f \bar{U}^2}{2}\right)\frac{L}{D}C_t \tag{5-10}$$

where $L = \dfrac{\pi}{4D}$

$$C_t = \frac{\text{solid volumetric flow rate}}{\text{total volumetric flow rate}}$$

From his experiments Schuchart also recommends a pressure-drop expression for a straight section of a pipe:

$$\Delta P_{\text{straight}} = \left(\frac{\rho_f \bar{U}^2}{2}\right)\left(\frac{L}{D}\right)\left(\frac{A}{K^2}\right)C_t \frac{D}{D_p}\frac{\bar{U}}{\bar{U}_p}\left(1 - \frac{\bar{U}_p}{\bar{U}}\right)^2 \psi' \tag{5-11}$$

where

$$\frac{\bar{U}_p}{\bar{U}} = \left[1 + \left(\frac{\rho_s}{\rho_f} - 1\right)^{2/3}\left(\frac{D_p}{D}\right)^{2/3}\left(1 + \frac{200}{Fr - Fr_0}\right)\right]^{-1} \tag{5-12}$$

$$\psi' = \frac{24}{Re} + \frac{4}{Re^{1/2}} + 0.4 \tag{5-13}$$

$$Re = \frac{(\bar{U} - \bar{U}_p)D_p}{\nu} \tag{5-14}$$

C, K, A = experimentally determined constants

Fr = Froude number $\dfrac{\bar{U}^2}{gD}$

Fr_0 = Froude number $\dfrac{\bar{U}_t^2}{gD}$

where $\bar{U}_t$ is the terminal velocity. The constant $A = 2.7$, C varies from 0.014 to 0.5 and K is given as 0.8 for glass spheres and 0.9 for plastic particles.

The Schuchart analysis has been applied to a coal flow situation in Example 5-4.

Example 5-4 Consider coal of density 1280 kg/m^3 flowing through a 0.0107-m-ID pipe through a 90° bend having a 0.305-m radius of curvature. The gas density is 11.8 kg/m^3 and the gas velocity is 10.0 m/sec. The volumetric solids concentration is 5% and the average particle diameter is 238 μm.

Equation 5-9 may be applied to yield

$$\Delta P_{\text{bend}} = \Delta P_{\text{straight}}\, 210\left(\frac{0.610\text{ m}}{0.0107\text{ m}}\right)^{-1.15}$$

$$= 2.01\Delta P_{\text{straight}}$$

The $\Delta P_{\text{straight}}$ must now be determined by Eq. 5-11. This entails calculation of all the parameters in this equation as given by Eq. 5-12 to 5-14.

$$Fr = 954$$

$$Fr_0 = 3.55 \text{ with } U_t = 0.610 \text{ m/sec from intermediate drag range}$$

$$\frac{D_p}{D} = 0.022$$

$$\frac{\bar{U}_p}{\bar{U}} = 0.967 \text{ with } C = 0.016$$

$$\psi' = 1.07$$

$$Re = 92.7$$

$$K = 0.8$$

$$C_t = 0.05$$

Thus, Eq. 5-11 yields

$$\Delta P_{\text{straight}} = 637 \text{ N/m}^2/\text{m}(L)$$

and for

$$L = \frac{1}{2} \pi 0.610) = 0.958 \text{ m}$$

$$\Delta P_{\text{straight}} = 610 \text{ N/m}^2$$

then

$$\Delta P_{\text{bend}} = 1.227 \text{ kN/m}^2$$

5-6 ACCELERATION LENGTH REPRESENTATIONS

As mentioned in the introduction to this chapter, the acceleration length could be employed to calculate the pressure loss around a bend. This procedure assumes that the acceleration length represents the largest contribution to the pressure loss around a bend. One can proceed in two ways to determine this acceleration length. Chapter four showed Yang's analysis for an acceleration length. This set of equations may also be applied here for the bend contribution to the overall pressure loss. In the dense-phase regime things may be simplified greatly by using Eqs. 4-14 and 4-15 for the frictional factor representation. The other technique is to use Rose and Duckworth's (1969) empirical acceleration length. Rose and Duckworth's equation for the pressure loss in acceleration length is

$$\Delta P_{\text{accel}} = 1.12 \left(\frac{\rho_f \bar{U}^2}{2} \right) \left(\frac{W_s}{W_g} \right) \phi_4 \phi_5 \qquad (5\text{-}15)$$

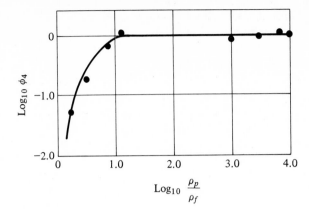

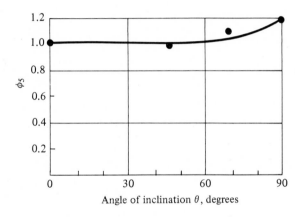

Figure 5-2 Calculation factors from Rose and Duckworth (1969). Reproduced with permission of *The Engineer,* Morgan-Grampean House.

Figure 5-3 Calculation factors from Rose and Duckworth (1969). Reproduced with permission of *The Engineer,* Morgan-Grampean House.

where ϕ_4 and ϕ_5 are found from Fig. 5-2 and 5-3. The acceleration length determination is also given by Rose and Duckworth:

$$L_{accel} = 6D \left[\frac{W_s}{\rho_f g^{1/2} D^{5/2}} \left(\frac{D}{D_p} \right)^{1/2} \left(\frac{\rho_p}{\rho_f} \right)^{1/2} \right]^{1/3} \qquad (5\text{-}16)$$

5-7 COMPARISON OF PRESSURE DROPS IN BENDS AND DESIGN RECOMMENDATIONS

Table 5-3 lists the pressure losses calculated for the various correlations and procedures suggested. It should be noted that if the piping systems have long sections of straight pipe, a few bends will do little to change the overall pressure loss in a system; however, if the piping arrangement has many bends and turns, pressure losses may well control the overall energy losses in the system.

Table 5-3 Pressure drop (kN/m^2) due to a 90° sharp bend for two-phase flow using different correlations

Coal flow, kg/sec	ΔP_{design}	$\Delta P_{Schuchart}$ 0.3048-m radius	$\Delta P_{Schuchart}$ Sharp bend	ΔP_{Yang} (accel)	$\Delta P_{Duckworth, Rose}$
0.0316	1.67	0.434	6.69	4.48	2.21
0.0341	1.78	0.465	7.24	4.76	2.38
0.0447	2.23	0.607	9.38	6.41	3.13
0.0485	2.39	0.655	10.13	7.03	3.40
0.0517	2.51	0.696	10.83	7.58	3.62
0.0592	2.79	0.793	12.3	8.69	4.15
0.0605	2.83	0.814	12.55	8.89	4.23
0.063	2.92	0.841	13.03	9.38	4.42
0.0687	3.09	0.910	14.20	10.27	4.81
0.0920	3.74	1.20	18.61	14.41	6.44
0.0964	3.85	1.25	19.51	15.31	6.75

From recent analysis of pressure drops in bends, the Schuchart model appears to give the most realistic representation of existing data and is therefore recommended in design calculations. If one encounters very fine particles, then this correlation must be modified in line with the data of Mason and Smith. A suggested format for the effect of particle size is

$$\frac{\Delta P_{bend}}{\Delta P_{straight}} = a \left(\frac{2R_B}{D}\right)^b \left(\frac{D_p}{D}\right)^c \tag{5-17}$$

In the case of fine powders the data all indicate that the pressure losses predicted by existing correlations overestimate the pressure losses that are actually measured. In some cases actual drag reduction may be present.

5-8 VELOCITY IN THE BEND

As the particle traverses a bend its velocity changes. Haag (1967) and Kovacs (1967) have analyzed the behavior of particles as they slow down in bends. The overall pressure loss in a bend may be made up of this deceleration of the particle in the bend plus the acceleration length term. Consider the horizontal bend shown in Fig. 5-4. The angle ϕ varies from 0 to 90° for a right-angle turn in the horizontal plane. A single particle momentum balance on this system can be written as

$$\frac{dU_p}{d\phi} = \frac{R_B}{m_p} \frac{\rho_f}{8} D_p \pi C_D \frac{(U_f - U_p)^2}{U_p} - \mu_s U_p \tag{5-18}$$

where m_p = mass of a single particle

R_B = radius of curvature

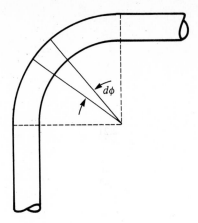

Figure 5-4 Differential section of a bend.

The last term in this expression represents the friction the particle sees while sliding along the outer surface of the bend. Since the centrifugal force tends to force particles to this outer surface, this is a reasonable representation of the actual situation. The term μ_s is the sliding frictional coefficient between the particles and the pipe wall. The condition of the pipe wall is a crucial parameter in this analysis. The rougher the surface, the greater the friction and the more reduction in the particle velocity. Haag has reported a value of 0.36 for μ_s for grain transport in a sheet metal duct. The pressure drop due to this particle velocity change can be written as

$$\Delta P = \frac{W_s}{A m_p} g \int_0^{\pi R_B/2} \frac{F}{U_p} dL \tag{5-19}$$

with the force term F written as

$$F = \frac{\rho_f}{2} (U_f - U_p)^2 \frac{\pi}{4} D_p^2 C_D - \frac{\mu_s m_p U_p^2}{g R_B} \tag{5-20}$$

The particle velocity reduction in a bend has been determined using Haag's analysis in Example 5-5.

Example 5-5 Consider the pneumatic flow of gas and solids in a horizontal bend. For analysis on the solid flow, consider that the fluid drag force is negligible in comparison to the sliding force of the particle on the bend surface (Haag, 1967). See Fig. 5-5.

A force balance on the particle is given as

$$m_p \frac{dU_p}{dt} = -\mu_s \mathbf{N} = -\mu_s m_p \frac{U_p^2}{R_B}$$

where $\mathbf{N}$ is the normal force.

Since

$$U_p \, dt = R_B \, d\phi$$

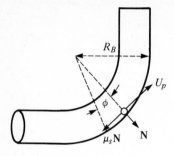

Figure 5-5 Forces on a particle in a bend. (Haag, 1967). Reproduced with permission of British Chemical Processing.

then

$$\frac{dU_p}{U_p} = -\mu_s d\phi$$

Upon integration with the limits $\phi = 0$ and $U_p = U_{p0}$, this gives

$$U_p = U_{p0} e^{-\mu_s \phi}$$

For the case where $\mu_s = 0.36$ and $\theta = \pi/2$,

$$\frac{U_p}{U_{p0}} = 0.57$$

which means that the particle has lost 43% of its velocity as it traverses the bend.

5-9 EROSION

Mason and Smith (1972) have found two weak points in the bend to be present in their analysis of erosion in bends. The erosion resistance of different materials changes as the angle of impingement changes. When the particles are harder than the bend material, rapid erosion of the bend occurs at low angles of impingement (15 to 39°). The rate of wear at this primary wear point decreases as the impact angle increases, until no measurable further increase in erosion is seen for impact angles between 80 and 90°.

Later studies by Mills and Mason (1977) considered the effect of particle concentration on erosion in bends. It was found by these investigators that the effect of particle concentration on erosion is slight, however, the depth of penetration of the particles into the bend surface material increases significantly with increase in particle concentration. The specific erosion measured as the mass of material eroded from the bend per unit mass of product conveyed is related to the solids loading as

$$\text{Specific erosion} = \text{constant} \times \left(\frac{W_s}{W_g}\right)^{-0.16} \tag{5-21}$$

The design of gas-solid transport systems thus would seem to indicate that a large radius of curvature should be used in order to cut down on the wear or erosion on the pipe wall. Another method of handling this situation is to use regular 90° elbows or bends on small R_B/r_0 ratios. When this is done, solid material builds up in the bend, forming a protective coating on the inside surface of the bend so that the flowing solid material erodes against itself and not the pipe wall. In addition, the use of a T connection with one leg plugged as a bend is an advantage. In such cases this leg will become impacted with solid material, quickly forming the erosion surface. Should an upset occur in the system, this leg could easily be uncapped and the line cleaned by ramming from outside or applying higher pressure from inside.

PROBLEMS

5-1 Gas at 1013 kN/m^2 pressure is flowing through a series of piping consisting of 15.2 m of horizontal distance and 6.10 m of vertical distance. The piping details include the following: two 45° elbows, four 90° square elbows, two gate valves, and one 180° close return bend. The gas velocity is 30.48 m/sec and the pipeline has a diameter of 0.0508 m. Consider the gas to be combustion products. What is the overall pressure drop for the system and what percentage of this drop is due to the above connections?

5-2 Consider the transport of carbon dioxide in a 0.0254-m-diameter pipe at a Reynolds number of 50,000. The pressure of the system is 203 kN/m^2 and the temperature is ambient. Determine the pressure drop through a bend for the following angles of inclination: 45°, 90°, and 180°. Assume the bend to be sharp.

5-3 What radius of curvature for a bend would be required in Prob. 5-2 for the 90° bend to have a pressure drop of 10% greater than an equivalent length of straight pipe?

5-4 Consider the transfer of 0.0252 kg/sec of 200-μm-diameter coal particles by carbon dioxide at an average gas velocity of 6.1 m/sec in a 0.0508-m-ID pipe. Determine the pressure loss around a 90° bend having a 0.61-m radius of curvature. The gas is at 3441 kN/m^2 and 38°C. Use the design suggestion in this chapter and Schuchart's analysis to determine the pressure loss. Compare the two techniques.

5-5 For the same conditions as stated in Prob. 5-4, analyze the acceleration length using Yang's analysis. Compare with the results of Prob. 5-4.

REFERENCES

Haag, A.: *Brit. Chem. Eng.* **12**:65 (1967)
Ikemori, K., and H. Munakata: *Pneumotransport* 2:A3–33 (1973).
Ito, H.: *Trans. ASME J. Basic Eng.* **81D**:123 (1959).
———: *Trans. ASME J. Basic Engr.* **82D**:131 (1960).
Kovacs, L.: *Pneumotransport* 1 (BHRA): C4 (Sept. 1967).
Mason, J. S., and B. V. Smith: *Powder Technol.* **6**:323 (1972).
———: *Pneumotransport* 2:A2–17 (1973).
Mills, D., and J. S. Mason: *Powder Technol.* **17**:37 (1977).
Morikawa, Y., K. Tsuji, K. Matsui, and Y. Jittani: *Int. J. Multiphase Flow* **4**:575 (1978).
Perry, J. H.: *Chemical Engineers' Handbook*, 3d.ed., McGraw-Hill, New York (1950)
Rose, H. E., and R. A. Duckworth: *The Engineer* 478 (Mar. 28, 1969).
Schuchart, P.: *Chem.-Ing. Technol.* **40** (21/22):1060 (1968).
Sproson, J. C., W. A. Gray, and J. Haynes: *Pneumotransport* 2:B2–15 (1973).

SIX

ELECTROSTATICS

6-1 INTRODUCTION

The phenomenon of electrification in gas-solid transport is often treated as some black magic or Maxwell's demon at work. To reinforce this black magic idea one only has to look at Wilcke's (1957) triboelectric series, which is a list of materials arranged in order so that if two of the materials are rubbed together, the high one in the series is positive and the lower one is negative:

Polished glass
Wool
Wood
Paper
Sealing wax
Unpolished glass
Lead
Sulfur
Other metals

Often, if an explanation cannot be found for an observation, electrostatics is named as the cause. Or, if someone wants to complicate a discussion of gas-solid flow, the issue of electrostatics is interjected. Thus, an attempt will be made here to classify and explain the observed behavior of systems having electrostatic contributions in order to dispel some of the aura of mysticism that surrounds this phenomenon.

The electrification of particles can be great in gas-solid systems, since only one unit of electronic charge per 10^5 surface atoms needs to be transferred to cause electrical breakdown of the surrounding air. These large charges on small particles are related to the large surface/volume ratio that is present for these fine materials. Whitman's (1926)

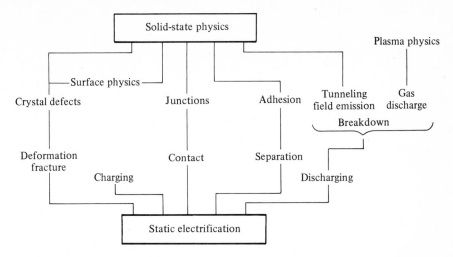

Figure 6-1 Interaction between solid-state physics and static electrification of solids (Krupp, 1971). Reproduced with permission of The Institute of Physics.

early experiments on blowing dust through a brass tube found charges on the dust of 10^{-6} C/kg. Often, these electrostatic effects are difficult if not impossible to duplicate, causing much frustration to the experimenter. If only certain things could be made constant, then the observations could be repeated. Parameters can be made constant, but the experimenter must be careful to overturn all aspects of the experiments to ensure identical experimental conditions. Some obvious examples of parameters of importance are the humidity and temperature of the gas, moisture content of the solids, surface condition of the flowing solids, and surface condition of the tube wall. Flow rates of the solid and gas stream and their loadings, particle size and shape, and overall chemistry are other important factors in electrostatics in gas-solid flow systems.

In this time of energy consciousness, an early study is of interest. Vollrath (1932) proposed static electrification by gas-solid flow as a high-voltage generator. The system he studied generated 260 kV at $\times 10^{-5}$ A. Such a system would yield 20.8 W of of power. In general, solid surface charges may be generated by contact and deformation and cleavage. Modern solid-state physics is related quite closely to charge generation.

The introduction to Chap. Two on energy bands in solids and their association to electrostatic adhesion should be referred to, since static electrification has the same bases as electrostatic adhesion forces. Krupp (1971) has given a convenient diagram showing the interactions between static electrification and solid-state physics. See Fig. 6-1.

6-2 EXPERIMENTS ON THE ELECTRIFICATION PROCESS IN SOLID-SOLID CONTACTING

As one goes through the literature of static electrification, contradictory results are seen throughout. These results point out the need for studies on materials having clean, smooth surfaces and on pure compounds with single crystals, if possible, in order to obtain reproducible results. Aqueous films and external fields only complicate the issue. To achieve clean surfaces, especially on glass, various cleaning techniques have been suggested. Generally, organic solvents leave residues on the surfaces. Peterson (1954) subjected glass surfaces to a series of acid washes followed by a distilled water bath with success. Once studies are performed under these exacting conditions, the parameters may be expanded to more complex systems.

The earliest and easiest technique for measuring static electrification consists of rubbing two surfaces together, separating them, and then measuring the accumulated charge with a Faraday cage. Measurements may also be done by charging dust, i.e., by blowing the dust on a surface connected to a Faraday cage.

Kunkel (1950) experimentally studied the dust-charging mechanism and made the following observations:*

1. Charging results from separation of surfaces in contact.
2. At less than 90% relative humidity, there was little effect of water films on charging.
3. Both homogeneous and heterogeneous charging can occur.
4. The average charge increases somewhat more slowly than the square of the radius of the particle.
5. Dissimilar substances can exchange charged carriers, electrons or ions.
6. Highly asymmetric charging of dust clouds on dispersion with surfaces leads to segregation of charge and ultimate sparks.
7. Substances that charge readily are (a) red lead with sulfur, (b) dry solid acids on metals (+), (c) dry solid alkalis on metals (−), (d) sulfur, (e) HgS on brass, (f) corn flour on brass, (g) charcoal on metal, (h) talc on metal, (i) $CaCO_3$ on metal, (j) sugar dust on glass (+) and sugar dust on copper (−), (k) coal on iron, (l) snow on ice, (m) CO_2/snow, (n) Hg on glass.

Another electrification technique involves rolling a ball of test material for study in a cylinder separated by an insulator. The two halves of the separated cylinder are then connected to an electrometer for measurement of the charge generated in the system. Peterson (1954) and Wagner (1954) have studied static electrification in this manner. Several observations may be made from these studies:

1. The maximum charge buildup on the surface can be represented by the total saturation charge.
2. The charging rate is governed by the rate of exposure of new uncharged contact area,

$$\frac{1}{q_s}\frac{dq}{dt} = \left(\frac{Mg}{E_1 R^5}\right)^{1/3}\frac{v_s}{10} \tag{6-1}$$

where E_1 = elastic modulus

v_s = rolling speed

q_s = saturation of sphere

R = radius of sphere

*List reprinted with permission from Loeb (1950), Springer-Verlag.

3. The saturation charge decreases as the surface finish becomes rougher. High saturation charging seems to be attributable to electrostatic forces.
4. The saturation charge is proportional to the contact potential.

See Example 6-1 for an application of Peterson and Wagner's formula for charging rates.

Example 6-1 From Peterson and Wagner's studies the charging rate on a particle can be related to its rolling speed. Consider a glass sphere and a plastic sphere of radius 2.5 mm traveling at 0.61 m/sec on a metal surface such as a cage. Determine the charging rates for these particles.

	Elastic modulus, N/m^2	ρ_p, kg/m^3
Glass	6.89×10^{10}	2,400
Plastic (nylon)	2.76×10^{10}	1,150

Since

$$\frac{1}{q_s}\frac{dq}{dt} = \left(\frac{Mg}{E_1 R^5}\right)^{1/3}\frac{v_s}{10}$$

then

$$\frac{1}{q_s}\left(\frac{dq}{dt}\right)_{glass} = \left[\left(\frac{1.57 \times 10^{-4}\text{ kg}}{6.89 \times 10^{10}\text{ N/m}^2}\right)\left(9.8\text{ m/sec}^2\frac{1}{0.0025^5\text{ m}^5}\right)\right]^{1/3}\frac{0.61\text{ m/sec}}{10}$$

$$= 3.73 \times 10^{-2}\text{ sec}^{-1}$$

$$\frac{1}{q_s}\left(\frac{dq}{dt}\right)_{nylon} = \left[\left(\frac{0.753 \times 10^{-4}\text{ kg}}{2.76 \times 10^{10}\text{ N/m}^2}\right)\left(9.8\text{ m/sec}^2\frac{1}{0.0025^5\text{ m}^5}\right)\right]^{1/3}\frac{0.61\text{ m/sec}}{10}$$

$$= 3.96 \times 10^{-2}\text{ sec}^{-1}$$

Ruckdeschel and Hunter (1975) have more recently analyzed contact electrification from a kinetic aspect. They considered the process a nonequilibrium one when the contacting occurs rapidly.

The dynamics of the electrification process can help distinguish between the possible mechanisms. Electron tunneling may take place up to a separation of 20 Å, while ions cannot tunnel through 1 Å. In general, Ruckdeschel and Hunter (1975) found that the speed of rotation of the particle in a cylinder and the outgassing have fundamental relationships to electrification. The charge-up time appears to be of geometrical origin and is related to the time required to contact the total surface of the bead at a given rolling speed. The gas pressure-dependence of the final charge separation appears to be determined by the atmosphere breakdown and thus is of secondary importance. The asymptotic separation charge q_m has been related to the separation velocities as

$$q_m = q_0\frac{v_s}{v_0}\left(1 + \frac{v_s}{v_0}\right)^{-1} \tag{6-2}$$

where q_0 = initial charge on the particle

v_0 = characteristic velocity

v_s = surface separation velocity or rolling speed

Example 6-2 gives a numerical value for the charge drainage.

Example 6-2 In a particular gas-solid flow system the particle velocity has been measured to be 9.1 m/sec. Ruckdeschel and Hunter have related to asymptotic separation charge to the surface separation velocity seen by the particle that bombards the walls of the pipe. If the turbulence ejection sequences noted by several investigators have been taken into account, the separation velocity of a particle bombarding the wall is approximated at 1.5 v_p. With this information, the asymptotic separation charge can be determined.

$$\frac{q_m}{q_0} = \frac{v_s}{v_0} \left(1 + \frac{v_s}{v_0}\right)^{-1}$$

assuming
$$v_0 = v_p = 9.1 \text{ m/sec}$$

$$v_s = 1.5(9.1) = 13.7 \text{ m/sec}$$

Thus,

$$\frac{q_m}{q_0} = 0.6$$

The asymptotic separation charge is 60% of the initial value.

The surface separation velocity is similar to the breakaway speed that Imianitov (1958) speaks of for particles colliding and generating an electrostatic charge.

6-3 BASIC CHARGE ANALYSIS

Before proceeding to the consideration of charging by gas-solid flow, it is advisable to review the basic physics of simple point charges. The basic unit of measuring charge is the coulomb (C). The electron, which is the fundamental charging unit, is measured with a charge of 1.60206×10^{-19} C. The total charge on a particle can be expressed as the product of the number of electrons times the charge of one unit. Current is the measure of the number of electrons or charges passing a fixed point over a certain period of time.

If one has a charge q_1, this charge sets up an electric field around itself and this field will act on any charge q_2 in its vicinity to produce a force F (see Fig. 6-2). The force on a charged particle is related to the electric field, which is the gradient of the electric potential. The force can be written as

$$F = Eq \tag{6-3}$$

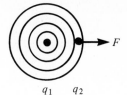

q_1 q_2 **Figure 6-2** Force on point charges.

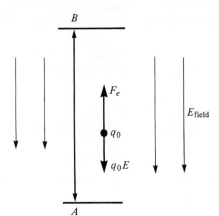

Figure 6-3 Point charge and electric field interaction (Resnick and Halliday, 1978). Reproduced with permission of John Wiley & Sons, Inc.

This expression may be likened to the gravitational field and the expression

$$F = gM \qquad (6\text{-}4)$$

For a whole field of charges similar to those present in gas-solid flow, a series of forces act on the particle producing the resulting force

$$F_1 = \sum_{i=1}^{n} F_{1i} \qquad (6\text{-}5)$$

A test charge or particle moving in a uniform electric field can be analyzed by a force balance. Figure 6-3 shows the representation of a test particle with a charge q_0 moving in a uniform electric field E by an external agent (conveying solids through a pipe). In this case, the electric force may be viewed as impeding the movement of the test particle from A to B. This viewpoint is a simplification of what actually happens in gas-solid transport systems. The resultant force is then

$$F = F_e - q_0 E \qquad (6\text{-}6)$$

where F_e = applied external force

6-4 DETAILS OF CHARGING BY GAS-SOLID FLOW

The basic charging of solid particles in a gas stream with a confining wall is assumed to depend on the contact potential V_c and the critical separation distance Z_0 (Baum, Cole, and Mobbs, 1970). The theory follows that of a parallel plate capacitor relating the charge to the potential by the capacitance:

$$q = C(V_c - V_1) = \frac{\kappa A}{Z_0}(V_c - V_1) \tag{6-7}$$

where V_1 = induced potential difference

$\quad A$ = area of an equivalent parallel plate

$\quad C$ = capacitance

$\quad \kappa$ = permittivity of air

Because of the charges induced on the wall, a potential difference between the wall and the particle results. Examination of Fig. 6-4 and 6-5 reveals that the potential on the particle surface is

$$V_y = \frac{q}{4\pi\kappa(2r-y)} - \frac{q}{4\pi\kappa y} \tag{6-8}$$

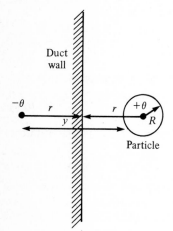

Figure 6-4 Model for potential difference between duct wall and particle (Baum, Cole, and Mobbs, 1970). Reproduced with permission of Institution of Mechanical Engineers.

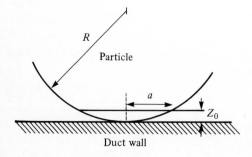

Figure 6-5 Definition of contact region. Reproduced with permission of Institution of Mechanical Engineers.

and the potential on the wall is

$$V_y = r = 0 \tag{6-9}$$

where q = total charge on a particle

The potential difference between the particle surface and the wall at a separation of Z_0 when $Z_0 \ll r$, is

$$V_1 = \frac{qZ_0}{2\pi R^2 \kappa} \tag{6-10}$$

The contact area is approximated to be

$$A \cong 2\pi R Z_0 \tag{6-11}$$

Loeb (1958) has suggested a correction factor on A as 10^{-7} to obtain the true contact area; thus,

$$A \cong 2\pi R Z_0 \times 10^{-7} \tag{6-12}$$

As the particles pass through the pipe, a field is set up that can further reduce the potential difference between the particles and the wall. For a uniform distribution of positively charged particles with a particle number density of N, the electric flux density at any radius is given as

$$\mathscr{D} = \frac{qNr}{2} \tag{6-13}$$

and the electric field strength may be written as

$$E = \frac{\mathscr{D}}{\kappa} = \frac{qNr}{2\kappa} \tag{6-14}$$

The potential difference between the wall and the distance Z_0 is

$$V_b = \int_{r-Z_0}^{r} \frac{qNr\,dr}{2\kappa} \cong \frac{qNrZ_0}{2\kappa} \tag{6-15}$$

The potential difference can now be written as

$$\Delta V = V_c - (V_1 + V_b) \tag{6-16}$$

During the charging mechanism of the solid with the wall there is a possibility of charge leakage. This leakage may generally be assumed to be due to conduction. With the charge leakage concept the charge added to a particle per collision with the wall can be written as

$$\frac{dq}{dn} = q_1 - K_1(q + q_0) \tag{6-17}$$

where $q_1 = \dfrac{\kappa_1 A \times 10^{-7} V_c}{Z_0}$

$$K_1 = \frac{A \times 10^{-7}}{2\pi R^2} + \frac{N(D/2)A \times 10^{-7}}{2} + l^*$$

l^* = charge leakage factor

q_0 = initial charge before collisions

The charge can then be found by integration of Eq. 6-17 as

$$q = \left(\frac{q_1}{K_1} - q_0\right)(1 - e^{-K_1 n}) \tag{6-18}$$

The total charge on a particle is then $Q = q + q_0$. From Eq. 6-18, the maximum charge on the particle can be written as $q_1/K_1 = q_{max}$. The value of K_1 can be expanded for spherical particles to give

$$K_1 = \frac{10^{-7}Z_0}{R}\left(1 + \frac{3W_s}{4\pi(D/2)R\rho_p U_p}\right) + \frac{T^*}{R^2} \tag{6-19}$$

with T^* a constant related to the charge leakage. The current is represented as the total charge added to all particles per second. The current then is

$$i = (\text{number of particles flowing/sec})\,q \tag{6-20}$$

The number of particles flowing per second is equal to

$$\frac{W_s}{(\pi/6)D_p^3 \rho_p}$$

giving the current as

$$i = \frac{W_s}{(\pi/6)D_p^3 \rho_p}\left(\frac{q_1}{K_1} - q_0\right)(1 - e^{-K_1 n}) \tag{6-21}$$

It should be noted that an equal but opposite charge takes place on the pipe. The number of collisions can be found by a few assumptions. If $B(W_s/W_g)$ represents the mass of particles striking a unit length per second, then the number of collisions incurred by one particle in a length L is

$$n = \left(B\frac{W_s}{W_q}\right)\left(\frac{1}{m_p}\right)\left(\frac{L}{W_s}m_p\right) = \frac{BL}{W_g} \tag{6-22}$$

Baum, Cole and Mobbs (1970) confirmed the linear relationship with the current and loading ratio or solids flow, as seen in Fig. 6-6. Using the above analysis, Examples 6-3 and 6-4 show the charge and current generated for flowing coal particles.

Example 6-3 The charge on a particle in a flow situation is dependent on the number of collisions it receives as it passes through a pipe. If the initial charge on a particle is zero, determine the behavior of the charge on the particle as the number of collisions varies. Consider the case of no charge leakage for a 50-μm-diameter coal particle in a gas-solid flow of a loading of 1.5 in a 0.0254-m-diameter pipe. The contact potential is 1 V. The gas velocity in the pipe is 30.5 m/sec.

The general equation for the charge can be written from Eq. 6-18 as

$$q = \frac{q_1}{K_1}(1 - e^{-K_1 n})$$

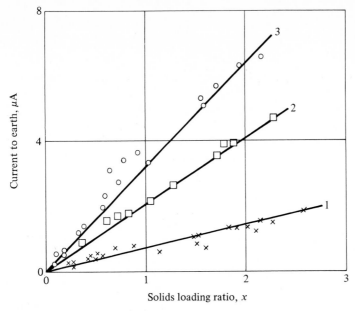

Figure 6-6 Effect of solid loading on current to earth from isolated section of brass pipe (Baum, Cole, and Mobbs, 1970). Air flow = 0.105 kg/sec, relative humidity = 24%, isolated length of pipe: 1, 2.74 m; 2, 11.28 m; 3, 27.7 m. (Reproduced with permission of The Institution of Mechanical Engineers.)

and

$$q_1 = \frac{\kappa A V_c \times 10^{-7}}{Z_0} \qquad K_1 = \frac{Z_0 \times 10^{-7}}{R} \left(1 + \frac{3W_g X}{4\pi r_0 R \rho_p U_p}\right)$$

The parameters of concern are

$$\kappa = \text{permittivity of air} = 8.9 \times 10^{-12} \, \text{C}^2/\text{N} \cdot \text{m}^2$$
$$Z_0 = 4 \, \text{Å}$$
$$V_c = 1 \, \text{V}$$
$$A = (2\pi)(25 \times 10^{-6})(4 \times 10^{-10}) = 6.28 \times 10^{-14} \, \text{m}^2$$
$$r_0 = 0.0127 \, \text{m}$$
$$R = 25 \times 10^{-6} \, \text{m}$$
$$\rho_p = 1200 \, \text{kg/m}^3$$
$$U_p \cong U_g = 30.5 \, \text{m/sec}$$
$$W_g = \text{gas flow} = 0.074 \, \text{kg/sec}$$
$$X = \text{loading} = 1.5$$

Thus,

$$q_1 = 1.40 \times 10^{-22} \, \text{C} \qquad K_1 = 0.525 \cdot 10^{-11}$$

and

$$q = 2.67 \times 10^{-11}[1 - \exp(0.525 \times 10^{-11}n)]$$

for

$$n = 10^4 \text{ collisions} \qquad q = 1.40 \times 10^{-18} \text{ C}$$

Example 6-4 Consider the case of gas-solid flow in a 0.025-m-diameter pipe. The number of collisions of the particles with a certain section of pipe is proportional to the length of the pipe and inversely proportional to the gas mass flow. Thus,

$$n = \frac{\alpha L}{W_g}$$

The current generated by this bombardment of particles with the wall is the product of the charge transferred on each times the number of particles flowing per unit time. Therefore,

$$i = \frac{W_s}{m_p} \left(\frac{q_1}{K_1}\right)\left[1 - \exp\left(\frac{-K_1 \alpha L}{W_g}\right)\right]$$

For a 3.66-m section of pipe with the same flow conditions as Example 6-5, determine the current generated. Assume $\alpha = 1$

In this case,

$$\frac{q_1}{K_1} = 2.67 \times 10^{-11}$$

$$K_1 = 0.525 \times 10^{-11}$$

$$L = 3.66 \text{ m}$$

$$W_g = 0.074 \text{ kg/sec}$$

$$\frac{W_s}{m_p} = \frac{X W_g}{(\pi/6)(50 \times 10^{-6})^3(1200 \text{ kg/ m}^3)} = \frac{1.5 \times 0.074}{7.85 \times 10^{-8}} = 1.41 \times 10^9 \text{ sec}^{-1}$$

Thus,

$$i = 9.76 \times 10^{-12} \text{ A}$$

Masuda, Komatsu, and Iinoya (1976) have also studied static electrification in solid-gas pipe flow, especially with respect to the wall collision. These investigators introduced a contact time concept in their analysis. This modification gives the current on the wall in a simplified form as

$$i \cong \frac{W_s \kappa V_c L \Delta t A_c}{Z_0 m_p}\left(\frac{\Delta n}{\Delta x}\right) \tag{6-23}$$

where Δt = contact time

A_c = contact area

n = number of collisions

L = length

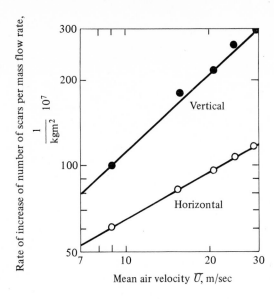

Figure 6-7 Relation between rate of increase of number of scars per mass flow rate versus the mean air velocity for quartz sand (Masuda, Komatsu, and Iinoya, 1976). Reproduced with permission of *AIChE Journal.*

The number of collisions with the wall were measured by Masuda, Komatsu, and Iinoya by analysis of the scars made by the impact of sand on a plastic film covering the inside of the tube. The rate of increase of the number of collisions per unit area was directly proportional to the solids flow rate and proportional to the gas velocity to a fractional power depending on whether the flow was horizontal or vertical:

	Power on $\bar{U}$
Horizontal flow	0.54
Vertical flow	0.93

For the average number of collisions of the particles through a unit pipe length, one finds the number to be proportional only to the velocity of the gas to the same power as previously discussed. The currents generated on the tube wall were also dependent on the solids flow rate as in Baum, Cole, and Mobbs' work, but a dependency on the mean velocity was seen to be a function of whether the test section was horizontal or vertical. Figure 6-7 shows this behavior clearly. A comparison between the Baum, Cole, and Mobbs analysis and the Komatsu, and Iionya analysis is shown in Example 6-5.

Example 6-5 The analysis of charge transfer by Masuda, Komatsu, and Iinoya (1976) utilizes a relaxation time and a time of contact in the development of charge and current generated by static electrification. This analysis may be compared to that of Baum, Cole, and Mobbs (1970) in that

$$K_{\text{BCM}} = \left(\frac{1}{N_0}\right)_{\text{MKI}} (1 - e^{-\Delta t/\tau})$$

where Δt = contact time

τ = relaxation time

$$(N_0)_{MKI} = \frac{\pi D_p^2}{2A_c} \Big/ \left(1 + \frac{3X\rho_f D\bar{U}}{4\rho_p D_p U_p^*}\right)$$

A_c = area of contact

X = loading ratio

Using this information in Example 6-3, the charge transfer can be calculated using the MKI analysis:

$$Q = \frac{q_1}{N_0}(1 - e^{-\Delta t/\tau})\left\{1 - \exp\left[\frac{-(1 - e^{-\Delta t/\tau})n}{N_0}\right]\right\}$$

Using $\tau = 4.5$ sec and $t = 10^{-6}$ sec, one finds

$$1 - e^{-\Delta t/\tau} = 2.22 \times 10^{-7}$$

This factor can have a profound influence on the number of collisions in order to generate the same current as in the case of Baum, Cole, and Mobbs.

6-5 CHARGE TRANSFER BY ANALOGY

The basics of charge transfer may also be presented in the form of two analogies. One involves using equations that describe the collision mechanics between particles and the wall, as presented by Timoshenko (1951) and developed by Soo (1967). This is quite similar to the basic heat-transfer analysis. The second approach is to use the penetration theory as given by Higbie (1935) and Danckwerts (1951) for heat, mass, and momentum transfer for the analysis of charge transfer.

Impact Analysis

The basic equation for charge transfer between two particles or a particle and a wall is not unlike the general transfer equations used for mass, heat, and momentum transfer. The rate of transfer of charge, which is electrical current, is proportional to the potential difference between the two surfaces, the driving force. Thus, one can write

$$\frac{dg}{dt} = i = A_c h \Delta V \tag{6-24}$$

Where A is the area of contact and h is the transfer coefficient. Both these terms can be analyzed in greater detail by relying on basic impact mechanics and conductance of charge. A detailed analysis of h and A has been given by Soo based on work by Timoshenko (1951) and Hertz (1881). The highlight of this development will be considered here, as well as calculations of these coefficients for some specific conditions. The charge-transfer coefficient is given as

$$h = \left(\frac{\sigma_1\sigma_2}{d_1 d_2}\right)^{1/2}\left[\left(\frac{\sigma_2 d_1}{\sigma_1 d_2}\right)^{1/2} + \left(\frac{\sigma_1 d_2}{\sigma_2 d_1}\right)^{1/2}\right]^{-1} \tag{6-25}$$

where σ is the conductivity of the species and d is the characteristic charge-transfer length. As can be seen, the charge-transfer coefficient depends intimately on the condition of the surface. Oxide films affect the conductance and transfer lengths. Adsorbed layers of water and, of course, any dirt in the system are very important effects to be considered in determining the conductance and charge-transfer lengths. The area term in Eq. 6-24 is difficult to measure. Some theoretical analysis of this quantity has been given by Timoshenko. Masuda, Komatsu, and Iinoya (1976) have some experimental evidence as to the size of these contact areas; sliding contacts were evident from the scars produced on plastic line walls of the transfer line. In his study of dispersed electrodes in a liquid flow system, Ross (1972) noted that the types of impacts the particles made with the electrodes were dependent on the loading ratio of the solid material. Dilute systems gave clean sharp hits of the particles with the wall, while more concentrated systems gave a sliding type of hit. These findings are undoubtedly also present in gas-solid systems, as evidenced by the contacting times and impulse measured in our experimental system (Weaver, 1979). Using Timoshenko's analysis for an elastic impact of a particle with the wall, the radius of contact is given as

$$A_c = 1.33 \left(\frac{FR}{E_i} \right)^{1/3} \tag{6-26}$$

where
F = contact force
R = radius of the particle
E_i = elastic constant = $\dfrac{1}{(1 - \mu_1^2)/E_1 + (1 - \mu_2^2)/E_2}$
μ_1, μ_2 = Poisson ratio
E_1, E_2 = elastic modulus

The contact pressure can be obtained from considering the momentum interchange between the particles and between the particles and the wall. Naturally, this quantity is then related to the velocity of the particle. The contact force is given as

$$F = \frac{4}{3\pi} \sqrt{\frac{R}{(k_1 + k_2)^2}} \left[\frac{5}{4} \frac{(1 + n^*)}{2} U_p^2 \cos\theta \, m_p \frac{3\pi}{4} \frac{k_1}{\sqrt{R}} \left(1 + \frac{k_2}{k_1} \right) \right]^{3/5} \tag{6-27}$$

where $k_1 = \dfrac{1 - \mu_1^2}{\pi E_1}$

$k_2 = \dfrac{1 - \mu_2^2}{\pi E_2}$

1 = particle
2 = wall
n^* = rebound coefficient
θ = angle of hit
m_p = mass of particle
U_p = particle velocity

Equation 6-27 can then be substituted into Eq. 6-26 to give the radius of contact:

$$a_c = 2^{4/15} R \left[\left(\frac{k_2}{k_1} \right)^{1/2} + \left(\frac{k_1}{k_2} \right)^{1/2} \right]^{1/5} N_{Im}^{1/5} \cos \theta^{1/5} \left(\frac{\rho_2}{\rho_1} \right)^{-1/10} \quad (6\text{-}28)$$

where $N_{Im} = 5\pi^2 U_p^2 \sqrt{\rho_1 \rho_2} \sqrt{k_1 k_2} \dfrac{1+n^*}{2}$

is the impact number. The contact area is then calculated as $A_c = \pi a_c^2$. Thus, from Eq. 6-24, if the contact potential ΔV is known, then h and A_c can be calculated from the preceding developments. In addition, the contact time over which the particle remains at the wall can be found as

$$\Delta t_c = \frac{2.94}{U_p \cos \theta} \left[\frac{5}{4} \frac{(1+n^*)}{2} U_p^2 \cos \theta \frac{3\pi}{4} \frac{m_p}{R^{1/2}} k_1 \left(1 + \frac{k_2}{k_1} \right) \right]^{2/5} \quad (6\text{-}29)$$

The previous development was based on the assumptions of spherical particles and elastic collisions. Bitter (1963) has relaxed these two assumptions to account for more general situations. For nonspherical particles Bitter modifies the density as

$$\bar{\rho}_p = \rho_p \left(\frac{D_p}{2r_0^*} \right)^3 = \rho_p \beta^{-3} \quad (6\text{-}30)$$

where D_p = diameter of a sphere having the same volume as the particle

r_0^* = radius of curvature of the particle surface

$\beta = \dfrac{2r_0^*}{D_p}$

Bitter has considered the possibility of collisions being plastic as well as elastic using the elastic deformation limit as a guide. Beyond a value of two-thirds of the yield pressure y^*, the collision has a plastic character. This gives the contact area for the elastic interaction as

$$A_{ce} = \pi r_0^{*2} (0.75 \times 3.02)^{2/3} \rho_p^{2/5} u_p^{4/5} \left(\frac{1-\mu_1^2}{E_1} + \frac{1-\mu_2^2}{E_2} \right)^{2/5} \quad (6\text{-}31)$$

For the plastic deformation, the area is given as

$$A_{cp} = \pi D_p^2 \sqrt{\frac{\rho_p}{3} \frac{r_0^*}{D_p}} y^* (u_p - v_e) \quad (6\text{-}32)$$

where v_e is the velocity of the particle at the elastic limit. This velocity is given as

$$v_e = 3.47 y^{*5/2} r_0^{*3/2} D_p^{-3/2} \rho_p^{-1/2} \left(\frac{1-\mu_1^2}{E_1} + \frac{1-\mu_2^2}{E_2} \right)^{-2} \quad (6\text{-}33)$$

Thus, the area for a plastic-elastic collision is given as the sum of Eqs. 6-31 and 6-32:

$$A_T = A_{ce} + A_{cp} \quad (6\text{-}34)$$

Table 6-1 Physical parameters for charge-transfer coefficients

Parameter	Soda glass	Copper	Aluminum	Plexiglas
Poisson ratio μ_i	0.23	0.36	0.33	0.35
Elastic modulus E, N/m^2	69×10^9	110×10^9	69×10^9	34.5×10^9
Conductivity σ_e, $1/\Omega \cdot$ m	10^{-18}	5.88×10^7	3.44×10^7	10^{-14}
Density ρ_p, kg/m^3	2,200	8,900	2,700	1,200

System	Transfer coefficient h, $1/\Omega \cdot$ m^2 (using contact distances of 4 Å)			
Glass/Cu	2.5×10^{-9}			
Glass/Plexiglas	2.5×10^{-9}			
Cu/Plexiglas	2.5×10^{-5}			

	Impact number $= \alpha v^2 (1 + r^*) \times 10^{-6}$ (α-values, m^2/sec^2)		
System	Glass	Copper	Plexiglas
Glass/Cu	0.1818	0.737	—
Plexiglas/glass	1.011	—	0.554
Cu/Plexiglas	—	3.10	0.419

Table 6-1 lists some physical parameters for a number of systems. These parameters should facilitate the use of the impact analysis.

Penetration Model

The impact analysis permits one to calculate transfer quantities using basic particle mechanics, a process that is not possible in considering molecular transfer of mass or heat in fluids flowing in a tube. Another approach to this charge-transfer mechanism is proposed by applying the penetration model. This model is probably more correct in regard to particle bombardment with the wall than when fluid eddies are considered. The single solid particle maintains its identity throughout the process. Considering the charge transferred on bombardment with the wall, one has, for an unsteady transfer process,

$$\frac{\rho_p}{m_p} \frac{\partial q}{\partial t} + \frac{\partial}{\partial x} \left(\frac{i}{A} \right) = 0 \tag{6-35}$$

Charge q and current i can be related to the potential as

$$q = CV$$
$$i = -\sigma_e A \frac{\partial V}{\partial x} \tag{6-36}$$

where $C =$ capacitance

$\sigma_e =$ conductivity

Using these definitions in Eq. 6-35, one has

$$\frac{\partial V}{\partial t} = \frac{\sigma_e m_p}{C \rho_p} \frac{\partial^2 V}{\partial x^2} \tag{6-37}$$

The grouping $(\sigma_e m_p / C \rho_p)$ is a diffusivity term. The x variable represents the distance into the wall. The current can be found from a solution of Eq. 6-36 with familiar boundary conditions:

$$V = V_0 \text{ for } t = 0 \quad \text{at all } x$$
$$V = V_1 \text{ for } t > 0 \quad \text{at } x = 0$$
$$V = V_0 \text{ for all times at } x = \infty$$

Various distributions of contact times of the solid particle with the wall could be considered. The exact distribution can be found from analysis of the experimental data. Considering a distribution of contact times of the solid particles with the wall as

$$\phi(t) = s e^{-st} \tag{6-38}$$

the electric flux can be determined as

$$\frac{i}{A} = -\sigma_e \left. \frac{\partial V}{\partial x} \right|_{x=0} \tag{6-39}$$

and

$$\frac{i}{A} = \int_0^\infty \sigma_e \, \phi(t) \Delta V \left(\frac{C \rho_p \pi t}{\sigma_e m_p} \right)^{1/2} dt \tag{6-40}$$

Integrating this expression gives

$$\frac{i}{A} = \Delta V \left(s \frac{C \rho_p}{\sigma_e m_p} \right)^{1/2} \tag{6-41}$$

which can be compared to Eq. 6-24 to show that

$$h = \left(s \frac{C \rho_p}{\sigma_e m_p} \right)^{1/2} \sigma_e \tag{6-42}$$

Some recent work in this area has shown the gamma distribution function to represent the particle hit data better than the exponential distribution function. Transfer coefficients for this penetration analysis are given in Example 6-6.

Example 6-6 Data taken on the bombardment of the wall of a Plexiglas tube show that the distribution of contact times can be represented as $\emptyset(t) = s e^{-st}$ with $s = 0.1 \text{ sec}^{-1}$. For 150-μm-diameter glass beads, estimate the transfer coefficient h for electrostatic charge transfer.

Considering the conductivity of Pyrex glass to be $0.012 \times 10^{-14} \ \Omega \cdot m^{-1}$ and the gap distance Z_0 on contact to be 10×10^{-10} m with a contact area A_c of 10^{-11} m^3, the capacitance equation is given as

$$C = 8.9 \times 10^{-12} \, C2/N \cdot m2 \left(\frac{A_c}{Z_0} \right)$$

$$C = 8.9 \times 10^{-12} \, C2/N \cdot m2 \left(\frac{10^{-11} \, m2}{10^{-9} \, m} \right)$$

$$C = 8.9 \times 10^{-14} \, C2/N \cdot m = 8.9 \times 10^{-14}$$

$(1F = 1 \, C/V)$. The diffusivity term $\sigma_e m_p / C \rho_p$ can now be found:

$$\left(\frac{\sigma_e m_p}{C \rho_p} \right) = (0.012 \times 10^{-14} \, \Omega \cdot m^{-1}) \, \frac{\pi}{6} \, \frac{(150 \times 10^{-6})^3 \, m3}{8.9 \times 10^{-14}} \, V/C$$

$$= 2.38 \times 10^{-15} \, m2/sec$$

The transfer coefficient h is then calculated as

$$h = [(0.1 \, sec^{-1})/(2.38 \times 10^{-15} \, m2/sec)]^{1/2} \cdot 0.012 \times 10^{-14} \, (ohm \, m)^{-1}$$

$$= 7.78 \times 10^{-10} \, (ohm \, m2)^{-1}$$

6-6 USEFULNESS OF ELECTROSTATIC CHARGING

Electrostatic charging in itself has not always been viewed as a detrimental effect. Cheng and Soo (1970) and King (1973) have done much to further the practical use of this phenomenon in measuring flow rates and thus giving a reliable flowmeter for two-phase flow. Cheng and Soo studied the electrostatic charging of ball and cylindrical probes in gas-solid flow; they derived the basic current equations and experimentally verified their findings. The net charge by impact of the particle with a ball probe is given by applying Eq. 6-24:

$$q_{net} \cong h(V_2 - V_1) A_{21} \Delta t \tag{6-43}$$

where h = charge-transfer coefficient from Eq. 6-25

A_{21} = area of contact

Δt = contact time

$V_2 - V_1$ = potential difference between particle and probe

The current generated by particle collisions is then

$$i = \frac{\eta_{21} A_p \rho_1 U_p \bar{q}_{21}}{m_p} \tag{6-44}$$

where η_{21} = ratio of the collision cross-sectional area of the particles to the projected area of the probe normal to the direction of flow

A_p = projected area of the probe

U_p = velocity of the particle

$\bar{q}_{21}$ = mean charge transferred per impact

m_p = mass of the particle

For a cylindrical probe the current is given as

$$i = \eta_{21} \frac{\pi}{2} N r_c l_c^2 \langle v_1^2 \rangle^{1/2} \frac{\bar{q}_{21}}{U_p}$$ (6-45)

where r_c = radius of the cylindrical probe

l_c = length of the probe

$\langle v_1^2 \rangle^{1/2}$ = velocity of the particles hitting the wall due to random motion

N = number density

Cheng, Tung, and Soo (1970) have used electrostatics to study the practical problem of measuring the flow rate of pulverized coal-gas suspensions. King (1973) also has used the electrostatic phenomenon as a measuring tool, but has looked at the unsteadiness of the signal employing a noise analysis to relate flow rates of gas-solid systems to this signal.

6-7 LESSENING ELECTROSTATIC CHARGING

In many cases electrostatics present very troublesome conditions in the handling of gas-solid systems. Often, elimination of the phenomenon would simplify system design considerably. The most common way to reduce the amount of charging that is transferred is to raise the relative humidity of the carrier gas. It has been seen experimentally that electrostatic effects can be totally eliminated for relative humidities greater than 75%. This was shown in two experiments performed by Peters (1970) on the flow of glass beads in a plastic tube (see Figs. 6-8 and 6-9). The pressure drop shown in these figures is attributable only to frictional effects, since gravity and acceleration effects were subtracted. The differences between the values of f_m/f_g in Figs. 6-8 and 6-9 can be attributed to electrostatic forces. This procedure of conditioning the air to eliminate charging is relatively simple, and it can be controlled closely. In most temperate climates, the dry winter air conditions generally set the stage for large electrostatic generation in gas-solid flows.

Two other techniques exist to reduce electrostatic charge in situ in a process by creating charge neutralization by ionization of a gas stream injection into the flow system. One technique involves the use of an ion gun first designed by Whitby (1961), and the other consists of passing a gas stream through a radioactive source to ionize the gas by a material such as polonium-210 or americium-241. The Whitby ion gun design is seen in Fig. 6-10. This gun consists of a fine wire arranged concentrically in a cylindrical tube or between two plates or rods. The positive/negative ion currents in the milliampere range flow between the generated corona and the electrode. It is desirable to operate the system at the highest practical ion output through the corona discharge without arcing. When an ac power supply is used, a stream of positive and negative ions is produced. The decay of the ion concentration once in the system tends to fall off rapidly. The concentration falls off as the reciprocal of the distance squared. Figure 6-11 shows a typical ion concentration with distance in a jet.

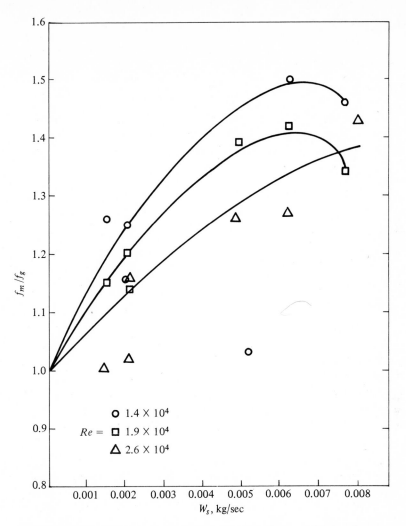

Figure 6-8 Ratio of two-phase friction factor to single-phase friction factor as a function of bead feed rate for 25-μm beads in silastic tube (Peters, 1971). Relative humidity = 25%, $f_m = (\Delta P/L)_f \, D/2\rho_m \bar{V}^2$, $\rho_m = (W_g + W_s)/[(W_g/\rho_f) + (W_s/\rho_s)]$, $f_g = (\Delta P/L)_g \, D/2\rho_f \bar{V}^2$. (Reproduced with permission of L. K. Peters)

The other technique using a radiation source consists generally of a cannister containing the alpha source. Polonium-210 is used by some manufacturers; the source has a strength of about 40 mCi. Air is blown through the source, picking up a stream of ions for charge equalization.

It should be noted that both the Whitby ion gun and the radioactive sources produce charge equilibrium over short distances, and care should be taken in applying these techniques to industrial situations.

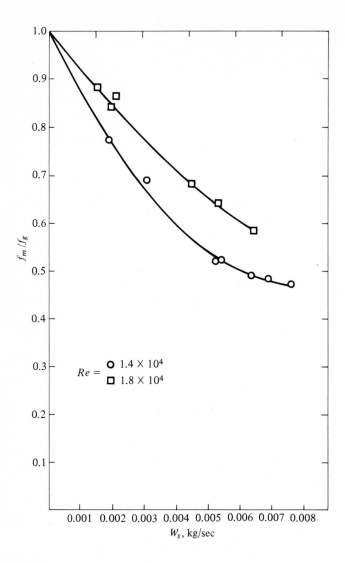

Figure 6-9 Ratio of two-phase friction factor to single-phase friction factor as a function of bead feed rate for 25-μm beads in silastic tube (Peters, 1971). Relative humidity = 80%. (Reprinted with permission of L. K. Peters.)

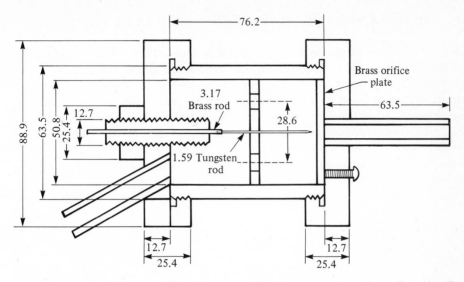

Figure 6-10 Ion gun (Peters, 1971). All parts are Plexiglas except electrode needle and orifice plate; all measurements are in millimeters. (Reproduced with permission of L. K. Peters.)

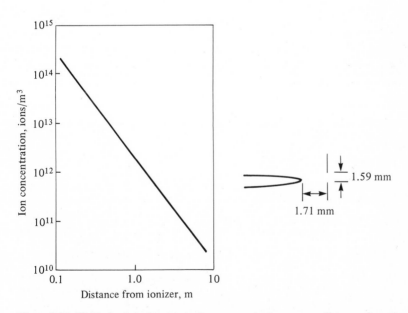

Figure 6-11 Whitby's data for jet ionizer concentration versus distance from ionizer (Whitby, 1961). Velocity $U = (7.75 \ \text{m}^2/\text{sec})(1/x)$, $p = 207 \ \text{kN/m}^2$, power = 3.5 kV ac. (Reproduced with permission of Scientific Instruments, American Institute of Physics.)

6-8 MEASUREMENT OF CHARGES ON PARTICLES

Min (1965) measured the charge on a particle in a gas-solid system by sampling the particles by means of a probe consisting of a stainless steel tube with a glass sleeve. The sampled particle interacts with the probe, which is connected to the grid of an electrometer tube creating a voltage pulse. In this manner the height and number of pulses can be measured. A spectrum of counts versus pulse height can be constructed as seen in Fig. 6-12. This technique permits one to discern the size of the average particle collision with the wall and the amount of charge transferred in the process. Min's apparatus was able to measure the total charge on the particle because of the capturing technique employed. For noncaptured particles, the total charge on the particle—if it is a nonconductor—is unlikely to be totally drained from the particle. A residue charge

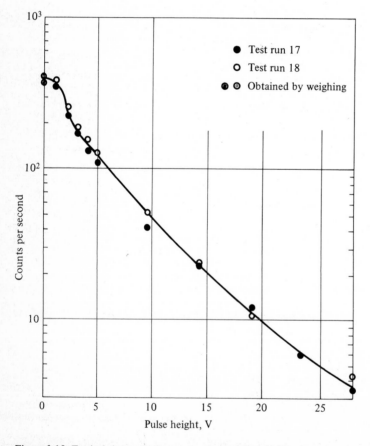

Figure 6-12 Typical integral charge spectrum (Min, 1965). Length of channel = 1.37 m, air velocity = 10.7 m/sec, particle feed rate = 0.000147 kg/sec. (Reproduced with permission of K. Min.)

will remain. The only sure method of obtaining the total charge is to capture the particle and drain the charge in a Faraday cage for measurement.

6-9 BASIC ENERGY LOSS IS IN GAS-SOLID FLOW WITH ELECTROSTATICS PRESENT

The measurement of energy losses by pressure drops in systems having electrostatics present is sparse. Often, increases in pressure drops are noted and it is tacitly assumed that this rise is due to static electrification effects. Richardson and McLeman (1960) and Clark et al. (1952) have made such observations. Quantitative measurements of pressure drops when electrostatics are present have been made by Peters, Bender, and Klinzing (1974) and Duckworth and Chan (1973). Baum, Cole, and Mobbs (1970) and Masuda, Komatsu, and Iinoya (1976) have measured the amount of charge generated on a pipe but have not determined the pressure drops present over this section of pipe. In an attempt to analyze the electrostatic effect on pressure drops the unified method of Yang has been modified to include the electrostatic force on the particle. Considering the force balance on a particle, one has

$$\Delta m_p \quad \frac{dU_p}{dt} \quad = dF_d \quad - \quad dF_g \quad - \quad dF_f \quad - \quad dF_{el} \tag{6-46}$$

mass   acceleration   drag   gravity   friction   electrostatic force

The drag, gravity, and friction terms are the same as shown in Chap. Four by Eqs. 4-2, 4-4, and 4-5. The electrostatic force can be written as

$$dF_e = E_x \frac{q}{m_p} (\Delta m)_p \tag{6-47}$$

where E_x = electric field

q = charge

m_p = mass

For the case of a nonaccelerated flow $\Delta m_p \, dU_p/dt = 0$, giving the particle velocity as

$$U_p = U_f - U_t \left(1 + \frac{f_p U_p^2}{2gD} + \frac{E_x q}{gm_p}\right)^{1/2} \cdot \epsilon^{2.35} \tag{6-48}$$

for vertical flow, and for horizontal flow,

$$U_p = U_f - U_t \left(\frac{f_p U_p^2}{2gD} + \frac{E_x q}{gm_p}\right)^{1/2} \cdot \epsilon^{2.35} \tag{6-49}$$

The acceleration lengths for systems involving electrostatics can now be calculated by using the accelerated flow case.

    In general, the overall pressure drop for such an electrostatic system can be measured. The analysis of this overall pressure drop can be interpreted in a similar manner to the "simple" two-phase flow case as

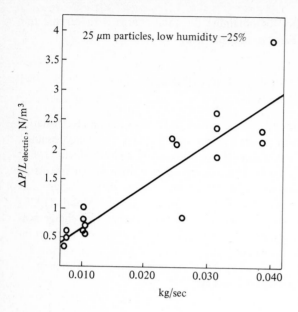

Figure 6-13 Electric pressure drop per unit length versus bead flow rate. Particles are 25 μm in diameter, humidity $\approx$ 25%.

$$\Delta P = \rho_p(1 - \epsilon)Lg + \frac{2f_g\rho_f U_f^2 L}{D} + \frac{2f_s\rho_p(1 - \epsilon)U_p^2 L}{D} + E_x \frac{q}{m_p}(1 - \epsilon)\rho_p L \quad (6\text{-}50)$$

The last term of this expression is attributable to the electrostatic effect.

Analyses have been performed on data taken in our laboratory on the effect of electrostatics on pressure drop (Weaver, 1979). Figure 6-13 is a plot of $(\Delta P/L)_{\text{electric}}$ produced in a system consisting of a Silastic tube with glass beads at low relative humidities. The $(\Delta P/L)_{\text{electric}}$ is the pressure drop measured over and above that predicted by Eq. 6-50, not including the last term (the electric effect).

As can be expected from these studies, the smaller the diameter of the flowing particle, the greater the electrostatic charge that is generated. The reduction in the particle size by a factor of two resulted in an increase in charge generated by nearly an order of magnitude.

6-10 DYNAMICS OF ELECTROSTATICS IN FLOW SYSTEMS

A particle in a flow field is influenced by the forces described in Chap. Three. In addition, when electrostatics is considered as a contribution to the force balance, one finds effects due to (a) net charge, (b) electric dipole, (c) electric field (due to charged particles or induced), and (d) magnetic dipole. For the case not having the magnetic effects, one can write these additional forces on the particle as

$$F_{\text{additional}} = q(E) + (\bar{p} \cdot E) \quad (6\text{-}51)$$

where q = charge on particle

E = electric field intensity

$\tilde{p}$ = dipole moment

For small or zero contribution from the dipole, this last term can also be dropped. Reviewing the basic dynamic equation of a single particle thus would add an extra *force qE* to the particle movement equation. Soo (1967)* has considered the motion of a particle where the dominant force on a particle is that due to the electrostatic force:

$$m_p \frac{dv_p}{dt} = qE \tag{6-52}$$

Soo has also analyzed the boundary-layer motion with electrostatic forces. Considering flow of uniformly charged particles over a conductive flat plate that is also charged, the equations for the fluid, particles, and potential V can be written as

$$\rho\left(u\,\frac{\partial u}{\partial x} + v\,\frac{\partial u}{\partial y}\right) = -\frac{\partial p}{\partial x} + F\rho_p(u - u_p) + u\,\frac{\partial^2 u}{\partial y^2} + E_x\,\rho_p\,\frac{q}{m_p} \tag{6-53}$$

and

$$u_p\,\frac{\partial u_p}{\partial x} + v_p\,\frac{\partial u_p}{\partial y} = F(u - u_p) + E_x\,\frac{q}{m_p} \tag{6-54}$$

and

$$\frac{\partial^2 V}{\partial x^2} + \frac{\partial^2 V}{\partial y^2} = -\rho_p\,\frac{(q/m_p)}{\kappa} \tag{6-55}$$

where E_x = electric field in the x direction

V = electric potential

An approximate solution to this system gives the boundary-layer thickness due to the particle as

$$\delta_p = \text{viscous contribution} + \text{electric field of plate} + \text{electric image force} \tag{6-56}$$

A dimensionless grouping arises out of the analysis comparing the viscous and electrostatic forces. The electroviscous number of the fluid is

$$N_{ev(\text{plate})} = \sqrt{\frac{\sigma x q}{\kappa m_p}}\,\frac{x}{\nu} \tag{6-57}$$

and for the particle,

$$N_{ev(\text{particle})} = \sqrt{\frac{\rho_p}{4\pi\kappa}\,\frac{q}{m_p}}\,\frac{x^2}{\nu} \tag{6-58}$$

where σ = surface charge density

x = distance

*Reproduced by permission of Ginn and Company.

For small electric effects,

$$(Re_x)^{5/2} \gg (Nev_{(plate)})^2 \tag{6-59}$$

$$(Re_x)^{5/2} \gg (Nev_{(particle)})^2 \tag{6-60}$$

For an air flow at STP with a charge/mass ratio of suspended particles of 10^{-4} C/kg and a velocity of 100 m/sec,

$$Re_x \approx 10^7 x$$

$$Nev_{(particle)} \approx 10^6 x^2$$

Thus, the conditions of Eqs. 6-59 and 6-60 are valid only near the leading edge of the plate.

For a charged plate with $\sigma = 10^6 \kappa$ and $E_y \big|_{y=0}$ 10^6 V/m,

$$Nev_{(wall)} \approx 10^6 x$$

$$Re_x \approx 10^7 x$$

PROBLEMS

6-1 In a gas-solid flow system of glass beads of average diameter 200 μm the transport gas velocity at STP is 30.5 m/sec with the relative humidity of the gas at 80%. The solids loading is 2.0 in a 0.0254-m-ID pipe made of Plexiglas. The gas relative humidity is changed to 40% and the pressure drop is found to increase by 25%. Determine the magnitude of the electrostatic forces present.

6-2 Using the same conditions as in Prob. 6-1, determine the particle velocity for the flow where electrostatics is present.

6-3 It has been suggested to use Soo's electrostatic cylindrical probe flow-measuring device to determine the mass flow of coal in a pipe. The pipe is 0.0254 m ID and the probe will be the same size. The average gas velocity is 24.4 m/sec. The probe length is 0.0254 m. The mean charge transferred per collision can be approximated at 0.5×10^{-15} C. The loading ratio is 1.5 and the average particle size of the solids is 50 μm. Determine the current generated in the probe at this operating condition.

6-4 In a given gas-solid flow arrangement, sand is being conveyed by air in a plastic pipe in the horizontal direction. The transfer of solids has been doubled. What would you expect as to the number of wall collisions experienced under these new conditions? Assuming electrostatics to play a major role, what kind of wall current would you expect on a metal pipe immediately following the plastic pipe? How would the loading affect the results?

6-5 A ceramic particle has a diameter of 100 μm and a charge of 10^{-15} C. An electric field of 10^6 N/C in the direction of flow is seen. Assuming steady state and the Stokes drag region to be valid, determine the particle velocity of a 9.1-m/sec vertical gas stream flow.

6-6 An electrostatic precipitator consists of two parallel plates separated by a distance of 0.06 m. The flow of gas through the system at 121°C and 207×10^3 N/m^2 pressure has a Reynolds number of 10^3. Combustion products are anticipated to have log normal distribution with the average log normal length diameter of 6.0 μm. The solid combustion product has a density of 1282 kg/m^3. The electric field is 10,000 N/C with a particle charge of 10^{-15} C. What is the drift velocity of the average particle, the length of the precipitator needed for capture of this particle, and the efficiency of the precipitator?

6-7 For Example 6-3 construct a plot of the maximum charge a particle can acquire q_{max} as a function of varying radius of a particle from 10 to 1000 μm.

6-8 Utilizing the data in Figs. 6-8 and 6-9 determine the electrostatic force contribution as a function of loading ratio for a Reynolds number of 1.9×10^4 and a 0.0254 m diameter tube with air at STP. What conclusions can be made?

6-9 Determine the transfer coefficient h as given by Eq. 6-25 for the binary flow systems listed below. Assume effect contact distance is 4 Å.

(*a*) Glass/Plexiglas
(*b*) Plexiglas/Plexiglas
(*c*) Glass/copper
(*d*) Copper/Plexiglas

Conductivity of glass is 10^{-18} $(\Omega \cdot m)^{-1}$, of Plexiglas is 10^{-14} $(\Omega \cdot m)^{-1}$, of copper is 5.88×10^7 $(\Omega \cdot m)^{/1}$.

REFERENCES

Baum, M. R., B. N. Cole, and F. R. Mobbs: *Paper 10, Proc. Inst. Mech. Eng.* 184 Pt C (1969–1970).

Bitter, J. G. A.: *Wear* 6(5):169 (1963).

Cheng, L., and S. L. Soo: *J. Appl. Phys.* 41:585 (1970).

Cheng, L., S. K. Tung, and S. L. Soo: *J. Eng. Power* 91:135 (Apr. 1970).

Clark, R. H., D. E. Charles, J. F. Richardson, and N. E. Newitt: *Trans. Inst. Chem. Eng.* 30:209 (1952).

Danckwerts, P. V.: *Ind. Eng. Chem.* 43:1460 (1951).

Duckworth, R. A. and T. K. Chan: *Pneumotransport* 2 (BHRA) A5-61 (1973).

Hertz, H.: *J. Math* (Crelle's J.) 92 (1881).

Higbie, R.: *Trans.* A.I.Ch.E. 31:365 (1935).

Imianitov, I. M.: *Sov. Phys. Dokl.* 3:815 (1958).

King, P. W.: *Pneumotransport* 2 (BHRA):D2 (1973).

Krupp, H.: *Static Electrification*, Institute of Physics, London, 1971.

Kunkel, W. B.: *J. Appl. Phys.* 21:829 (1950).

Loeb, L. B.: *Static Electrification*, Springer, Berlin, 1958.

Lyon, L. E.: *Physics and Chemistry of the Organic Solid State,* Vol. 1, Interscience (Wiley), New York, 1963.

Masuda, H., T. Komatsu, and K. Iinoya: *AIChE J.* 22:558 (1976).

Min, K.: "Particle Transport and Heat Transfer in Gas-Solid Suspension Flow Under the Influence of an Electric Field," Ph.D. Thesis, U. Illinois, Urbana, 1965.

Peters, L. K.: "A Study of Two-Phase Solid-in-Air Flow Through Rigid and Compliant Wall Tubes," Ph.D. Thesis, University of Pittsburgh, 1971.

——, D. W. Bender, and G. E. Klinzing: *AIChE J.* 20:660 (1974).

Peterson, J. W.: *J. Appl. Phys.* 25:907 (1954).

Resnick, R., and D. Halliday: *Physics for Students of Science and Engineering,* Part II, 3d ed. Wiley, New York, 1978, p. 627.

Richardson, J. F., and M. McLeman: *Trans. Inst. Chem. Eng.* 38:257 (1960).

Ross, T. K., and H. J. Wood: *Corrosion Sci.* 12:383 (1972).

Ruckdeschel, F. R., and L. P. Hunter: *J. Appl. Phys.* 46:4416 (1975).

Soo, L. L.: *Fluid Dynamics of Multiphase Systems*, Blaisdell, Waltham, Mass., 1967.

Timoshenko, S.: *Theory of Elasticity*, McGraw-Hill, New York, 1934, p. 339.

Vollrath, R. E.: *Phys. Rev.* 42:298 (1932).

Wagner, P. E.: *J. Appl. Phys.* 25:907 (1954).

Weaver, M. L.: "Electrostatics in Pneumatic Transports," M.S. Thesis, University of Pittsburgh, 1979.

Whitby, K. T.: *Rev. Scientific Instruments* 32:1351 (1961).

Whitman, V. E.: *Rev.* 28:1287 (1926).

SEVEN

MEASUREMENTS AND INSTRUMENTATION

7-1 INTRODUCTION

A crucial issue in gas-solid flow is still the correct way to measure the amounts of the gas and solids and the conditions of flow. Many advances have been made in instrumentation and the new instruments have shown much promise. Single-phase flow measurements are fairly well standardized except for a few unique applications, one being very, very slow flow. In the gas-solid flow field, inserting a measuring device in the line is generally unacceptable since abrasion will probably destroy the device. Also, if the line is small enough, plugging can easily result at that point. Generally, when devices for measurement are required to be attached to the pipe through an opening in the wall, the solid material has a way of migrating to the instrument sensor even if some precautions are made to eliminate this migration. Metal filters are finding increased application in this field to eliminate solid contamination of the sensor surface, since at times fine powders seem to violate even gravity effects. Nonobstrusive measurements of flow have ever increasing application in the gas-solid flow area. Light, sonic, radiation, and electrostatic fields are being applied in commercial metering devices.

7-2 STANDARD INSTRUMENTATION

By standard instrumentation is meant the standard equipment used for single-phase flow studies—manometers, orifice meters, venturis, etc. The manometer principle for measuring pressure drops is a very viable technique in gas-solid flow. Calibration of the

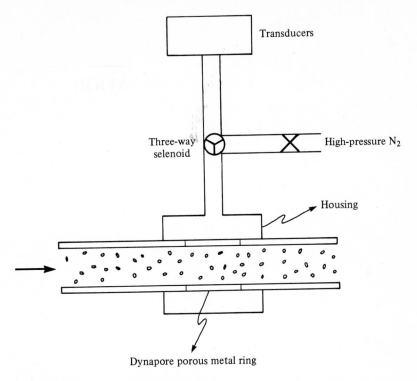

Transducers

Three-way selenoid

High-pressure N$_2$

Housing

Dynapore porous metal ring

Figure 7-1 Metal filter arrangement in pipe.

pressure drops with solid flow rates can be done, or for dilute-phase conditions, some of the existing correlations can be tested. Care should be taken to prevent plugging of the small-diameter manometer lines with solids. The use of a continual purge of gas through these lines is one suggestion. Another would be to employ the metal filter arrangement seen in Fig. 7-1, where the tube wall is made up of a metal filter. These metal filters are effective in preventing fines from migrating into the manometer lines, but they also can clog; therefore, a back-pressure device should be incorporated to blow back the filter if plugging occurs.

The standard manometer or micromanometer fluid arrangement can be employed or transducers can be attached at the ends of the pressure lead lines. The transducers can be equipped with strip chart recorder or digital outputs.

In dilute-phase regimes the principle of using a venturi and orifice meter in series (the Graczyk or BCR meter) can be employed. In this arrangement the venturi meter will measure the acceleration of the solids and the orifice meter will measure only the gas flow, since with closely placed pressure tops the particles will undergo little acceleration. This can be a relatively inexpensive solution to a gas-solid flow problem, especially if a venturi is already on hand (see Fig. 7-2). For dense-phase gas-solid flow or systems having sizable upsets, the orifice meter–venturi setup is subject to frequent plugging with solids. The unit is also applicable only for small-diameter lines.

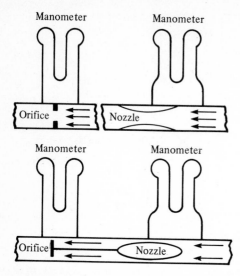

Figure 7-2 BCR gas-solid flowmeter (Saltsman, 1965). Reproduced with permission of Bituminous Coal Research, Inc.

7-3 LIGHT TECHNIQUES

When considering noninterfering techniques for measurement within a process, light techniques come to mind immediately. Much work has taken place in this area in the last several years with the many laser applications. Photomultiplier applications have been available even longer, but their refinement into a sensitive meter for gas-solid flow measurements was generally not advanced. In the dilute-phase transfer a photomultiple arrangement with a sodium lamp source can be applied. Again, calibration is necessary.

A concern in all light applications to gas-solid flow measurements is the cleanliness of the glass viewing port on the flow channel. Generally, if a solid is flowing in a metal tube, insertion of a glass section of tubing will cause electrostatics to develop. This causes deposits of the solids on the glass tube, thus cutting down the transmitted light signal. Some investigators have applied a gas jet directed at the glass surface as a method of keeping the viewing surface free of solid deposits. This technique may achieve this purpose, but it also changes the basic flow field so that the measurement of solids flow obtained is questionable. Some studies (Klinzing, 1980) have shown an equilibrium deposit to develop on the glass surface; this cuts down the signal by a fixed amount which appears to be reproducible over a number of conditions. The deposition of solids on the glass is dependent on a number of factors: electrostatics, humidity, solids properties, etc. If the constancy of some of these parameters is relatively certain, a constant number of particles will be deposited on the glass and this will not change over the period of operation of the process.

Lasers are power tools in gas-solids flows; they can be used in dilute-phase systems to measure the particle velocities. These laser Doppler velocimeters (LDVs) require adaptation of the flow arrangement and are still fairly expensive. The laser applied to gas-solid flow operates on the principle of shifting the frequency on the laser monochromatic light beam. In this instrument the Doppler shift of laser lights scattered from moving particles is detected by optical mixing of the scattered radiation with a reference beam from the same laser. The resultant heterodyne or beat signal is a measure of the velocity of the moving particle. The LDV has the ability to measure the point velocity of the flow with a high degree of accuracy without perturbing the flow. The Doppler shift ν_D is given by

$$\nu_D = \frac{n_R \nu_p}{\lambda_0}(\mathbf{r}_s - \mathbf{r}_i)$$

where λ_0 = vacuum wavelength of light

n_R = refractive index of medium surrounding the light

ν_p = velocity of particle

$\mathbf{r}_s, \mathbf{r}_i$ = unit vectors defining the direction of the scattered and incident radiation

The detection of ν_D can be accomplished by means of optical heterodyning of the laser light on a photomultiplier tube. The minimum sampling volume for a helium-neon laser is about 10^{-16} m^3. A number of investigators have used LDVs in studies on gas-solid systems (Dunning and Angus, 1967; Kolansky, 1976; Morrow and Angus, 1968; Peskin, 1973; Zlabin and Rozenshtein, 1975).

Recently, the LDV has uncovered some rather interesting information about gas-solid flow. The particle velocity intensity is seen to be greater than that of the gas velocity intensity at the center of the duct. Kane (1973) and Peskin (1973) have numerically simulated this same condition and have found that in high-shear regions the particle diffusivities are greater than the eddy diffusivities of the fluid. Kolansky (1976) has found similar results in experiments using the LDV. In addition, the smaller particles are found to fill the flow channel uniformly, while the larger particles go toward the outer boundary of the duct. The use of the LDV near a solid boundary with submicron particles has been noted by Kane, Weinbaum, and Pfeffer (1973) and Berman and Dunning (1973). The problem of gradient broadening, which is caused by having a finite measuring volume in regions of high-velocity gradient is a serious one. Figure 7-3 shows a typical laser arrangement in a flow measuring device.

A halogen lamp op/amp arrangement has been applied to the study of flow in dilute- and dense-phase gas-solid flows (Klinzing, 1980). This inexpensive equipment appears to work well over a wide range of solids loadings. Figures 7-4 and 7-5 show some of the details of this instrument installation with its associated electronics for the operational amplifier. Since this unit is small, it can be mounted easily on the viewing part of the flow system. This unit, like other units that depend on light transmission, obeys Beer's law of transmission which is logarithm-dependent. This behavior can cause some problems in accuracy, since a large change in loading can yield a small change in response for the illumination. See Example 7-1 for more actual data on this device.

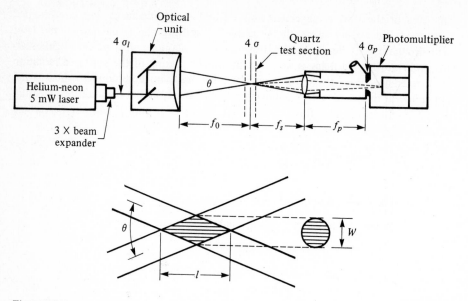

Figure 7-3 Laser arrangement (Kolansky, Weinbaum, and Pfeffer, 1976). Here, $f_0 = 130$ mm, $f_s = 41.7$ mm, $f_p = 105$ mm, $\lambda = 6328$ Å. (Reproduced with permission of British Research Association Fluid Engr.)

Example 7-1 A series of tests has been conducted to measure the solids flow in a pipe by use of the light-sensing device shown in Fig. 7-5. As coal flows with air through the glass section of pipe where the light-sensing arrangement is installed, a reduced signal is noted with increasing coal flow rates. Beer's law of illumination is used to analyze the light responses. The data are listed in Table 7-1. Figure 7-6 shows a plot of the data, light versus loading response.

Table 7-1

Run number	Coal flow, kg/m	Light response, $-\ln R/R_0$	Loading, kg coal/kg air
D1	0.0215	0.0993	3.12
D2	0.022	0.147	3.21
D3	0.0237	0.124	3.44
D4	0.0099	0.104	1.45
D6	0.0166	0.116	3.01
D7	0.0161	0.120	2.91
D8	0.0147	0.144	3.60
D9	0.0072	0.028	1.05
D10	0.0116	0.059	1.68
D11	0.0147	0.064	2.15
D12	0.0176	0.094	2.57
D13	0.0203	0.104	2.96

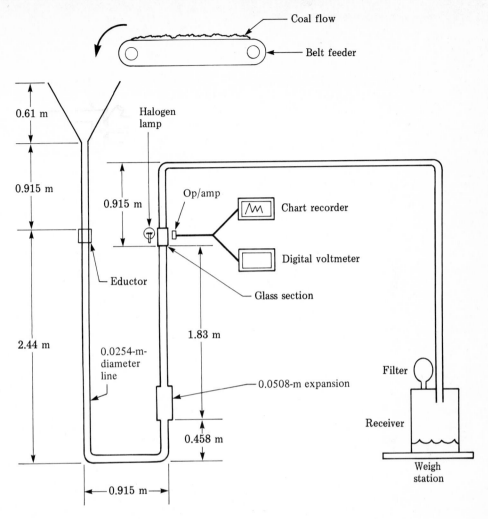

Figure 7-4 Test flow facility.

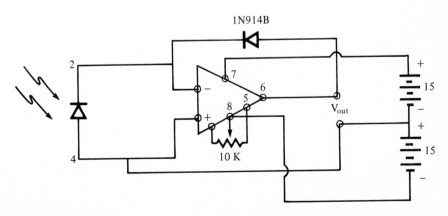

Figure 7-5 Op/amp diagram, United Technology with 450 photodiode.

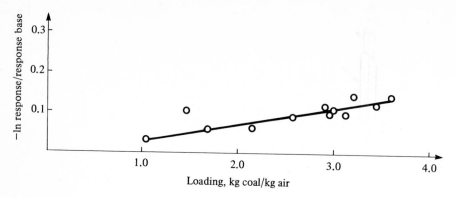

Figure 7-6 Light response ratio versus solid loading.

7-4 ELECTRO MEASUREMENTS

Electromeasurements rely on self-generated electroresponses, electrostatics, and imposed electric field applications. Generally, the electrostatic aspects have been tested more than the electric field applications, although an exciting new instrument employs rapidly changing electric fields to measure the changes in the dielectric properties of the flowing material. Cheng, Tung, and Soo (1970) have tested the principle of electrostatic generation in coal transport systems. Predominantly, they studied a probe insert in the flow, but in addition, some work was done on the charge generated on an electrically insulated section of pipe. Sizable electrostatic charges can be built up on flowing gas-solid systems, especially at low relative humidities of the gas stream. Figures 7-7 and 7-8 show a typical result of probe current generated versus the solids mass flow and diagrams of the probes themselves (Cheng and Soo, 1970). King (1973) also investigated the electrostatics generated by pin electrodes perpendicular to the flow or an electrode on the wall of a pipe. He correlated the unsteady electrostatic signal generated with the flow rate using random signal analysis; Fig. 7-9 shows a result of this work. Chao, Perez-Blanco, Saunders, and Soo (1979) made some modifications in this probe so that they could count the particle hits using an electronic counter. This technique was found to be superior to the older design.

Babh and Etkin (1974) have applied voltages to electrodes in flowing gas-solid systems in order to measure the changing electrical conductivity with mass flow rates. The probes were placed some 2.5 mm apart. As with anything placed in a gas-solid flowing system, erosion and clogging are problems.

Electrostatic and electrical conductivity methods are very dependent on the condition of both the solids and the gas. If uniformity in these parameters can be guaranteed, then these electrical methods should work well. A change in the relative humidity or the type and uniformity of the solids flowing can upset any correlation that may have been performed on a particular flow system.

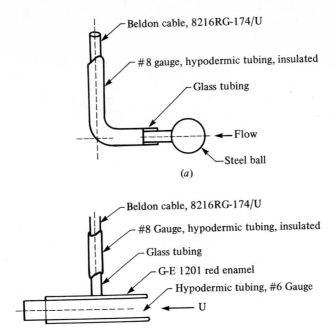

Figure 7-7 (*a*) Electrostatic ball and (*b*) cylindrical probes (Cheng and Soo, 1970).

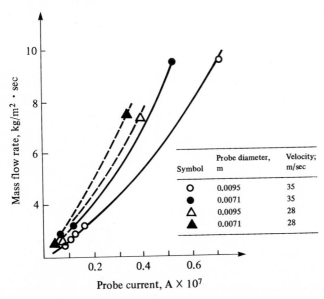

Figure 7-8 Electrostatic ball probe current at center of duct; effect of probe diameter and flow velocity at floating probe voltage (Cheng and Soo, 1969). Reproduced with permission of American Institute of Physics.

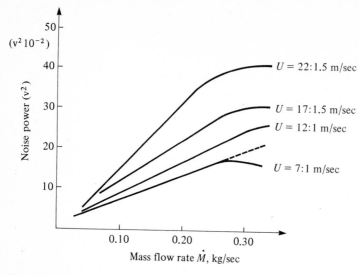

Figure 7-9 Linear plots of measured noise power against PVC mass flow at specified values of mean transport velocity (King, 1973). Reproduced with permission of BHRA Fluid Engr.

7-5 OTHER MEASUREMENT DEVICES

Bubble Injection

In a gas-solid system the term *bubble injection* may seem strange, but it is the same physical designation that is given to a bubble in a fluidized bed system containing solids and a gas. Injection of this bubble from a high-pressure source such as a tank of nitrogen through a solenoid valve is possible. This slug of gas can be used as a tracer to measure the solid velocity. This can be achieved by noting the passage of the gas slug with its higher-pressure wave flowing between two designated points where pressure sensors are situated. McLeman and Richardson (1960) have utilized this technique in their studies.

Tracer Gases

A tracer gas can also be injected into the flowing gas-solid stream in order to study some of the details of the flow process. By watching the diffusion or dispersion mechanisms of tracer gases such as helium, the turbulent diffusivity of the gas-solid system can be found. This adds considerably to the understanding of the transport process in gas-solid flows. In addition to gas tracers, heat sources can be inserted in the flow and their dispersive behavior noted. The work of Bobkowicz and Gauvin (1965) is notable in this area.

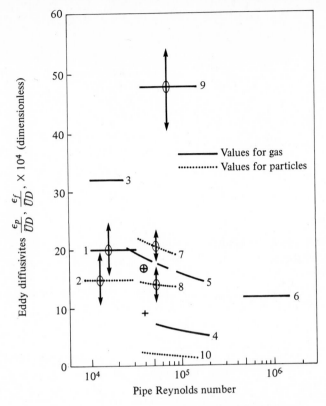

Figure 7-10 Collected values of average eddy diffusivity in pipe center. Arrows indicate spread of data; curves 1, 2, van Zoorev (1962); curve 3, McCarter, Stutzman, and Koch (1949); curve 4, Schlinger and Sage (1953); curve 5, Towle and Sherwood (1939); curve 6, Baldwin and Mickelson (1963); curves 7, 8, Boothroyd (1967); curves 9, 10, Soo, Ihrig, and El Kouch (1960); ⊕ (gas), Briller and Robinson (1969); + (fine particles), Boothroyd (1971). Reproduced with permission of Taylor and Frances, Ltd.

Figure 7-10 shows a compilation of data by various investigators on the eddy diffusivity for a gas-solid system.

Double Solenoid Isolation

One of the most difficult tasks in gas-solid flow is to determine the voidage present at a certain flow condition. From the voidage, the solid velocity can be found; thus, it is in a sense, the problem of finding the solid velocity. The use of a section of pipe equipped with two fast-action solenoid valves at given distances apart has been successfully applied in gas-solid studies. At prescribed flow conditions the valves are actuated simultaneously, isolating the section of pipe and trapping the solids present. The solids are then removed from the section and their volume and weight can be found for voidage and solid velocity determinations.

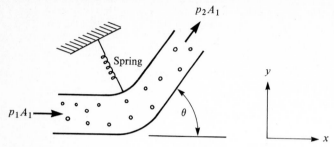

Figure 7-11 Elbow meter.

Force Instrumented Elbow

The use of a common elbow can be expanded to a solids flow measuring device. As the solids traverse an elbow, the elbow surface is bombarded by the solids, creating a considerable stress or force on this section. The flow can be equipped with a strain gauge arrangement and vibrationally isolated from the rest of the piping system (see Fig. 7-11). A force balance on the elbow can then relate the force on the strain gauge to the solids flowing inside the pipe. See Example 7-2 for details.

> **Example 7-2** Consider the details of the forces imparted to a bend as a gas-solid flow passes through it. The surface force on the pipe wall in the x direction can be written as
>
> $$F_x = p_1 A_1 - p_2 A_1 \cos \theta$$
>
> and for the y direction,
>
> $$F_y = -p_2 A_1 \sin \theta$$
>
> The momentum in these two directions can now be expressed as
>
> $$\text{Momentum}_x = \dot{M}\bar{V}_1 - \dot{M}\bar{V}_2 \cos \theta$$
>
> $$\text{Momentum}_y = 0 - \dot{M}\bar{V}_2 \sin \theta$$
>
> $$F_{x(\text{total})} = p_1 A_1 - p_2 A_1 \cos \theta + \dot{M}\bar{V}_1 - \dot{M}\bar{V}_2 \cos \theta$$
>
> $$F_{y(\text{total})} = \dot{M}\bar{V}_2 \sin \theta + p_2 A_1 \sin \theta$$
>
> The force acting on the spring at 45° is
>
> $$F = \sqrt{F_{x(\text{total})}^2 + F_{y(\text{total})}^2}$$
>
> If the angle $\theta = 90°$,
>
> $$F = \sqrt{(p_1 A_1 + \dot{M}\bar{V}_1)^2 + (p_2 A_1 + \dot{M}\bar{V}_2)^2}$$
>
> $$\dot{M}\bar{V} = \underset{\text{gas}}{\dot{M}_g \bar{V}_g} + \underset{\text{solids}}{\dot{M}_p \bar{V}_p}$$

Some simplifying assumptions permit calculation of the magnitude of the force on the spring. For $p_1A_1 \approx p_2A_1$ and $\dot{M}_p/\dot{M}_g = 3.0$ with a gas velocity of 45.7 m/sec at atmospheric pressure and 21°C, one finds for a 0.0254-m-diameter pipe,

$$p_1A_1 = (101.3 \times 10^3 \text{ N/m}^2)\left(\frac{\pi}{4}\right)(0.0254)^2 \text{ m}^2 = 51.3 \text{ N}$$

$$\dot{M}_g = \rho \bar{V}A = (1.2 \text{ kg/m}^3)(45.7 \text{ m/sec})\left(\frac{\pi}{4}\right)(0.0254)^2 \text{ m}^2 = 0.0278 \text{ kg/sec}$$

Thus,
$$\dot{M}_g = 3\ (0.0278) = 0.0834 \text{ kg/sec}$$

If the particle size of the solids is in the range of about 5 μm, then $V_g \cong V_p$; therefore,

$$(\dot{M}\bar{V}) = (0.0278 + 0.0834 \text{ kg/sec})(45.7 \text{ m/sec}) = 5.08 \text{ N}$$

Thus,
$$F = 51.3 + 5.08 = 56.38 \text{ N}$$

Screw Feeders

Screw feeders provide a convenient way to deliver solids to a gas-solid transport system or process, and they can also be used as metering devices. These units permit accurate delivery of precise amounts of solids. Generally, these units do not work on high-pressure systems but can work adequately up to about 2 atm pressure.

Piezoelectric Transducer

Mann and Crosby (1977) have employed a piezoelectric transducer for flow measurements in gas-solid flows. The principle of operation is based on the impact of individual particles with the transducer. This impact is recorded and the number of collisions per unit time can be transformed to a local particle velocity and solids flow rate (see Fig. 7-12).

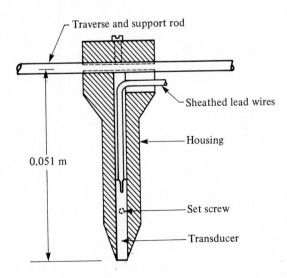

Figure 7-12 Piezoelectric detecting probe. From Mann, U. and E. J. Crosby: *IEC Process Des. Dev.* **16**:9 (1977). Reproduced with permission of American Chemical Society.

Cantilevered Beam

Gibson, Abel, and Fasching (1963) have inserted an instrumented cantilevered beam into the gas-solid flow to measure the flow rate of solids. The force produced on the beam is proportional to the amount of solids flowing. This unit naturally will suffer from abrasion over long periods of time.

Sonic Devices

The Argonne National Laboratory project (LeSage and O'Fallon, 1977) is developing a meter dependent on the attenuation of sound by the particles suspended in a gas. The path length, frequency, and power level of the acoustic are essential parameters to be set.

Capacitance Measurements

Another flow measurement device is based on the changing dielectric property of varying solids concentration in the flow path. One commercial unit operating on the capacitance principle is the Auburn meter. This unit is able to detect the flow of materials having different dielectric constants. Six sensing capacitors are located around a circular pipe cross section. The field is rotated at about 1000 rps. Figure 7-13 shows a diagram of the operating principle of the Auburn meter. The average dielectric constant of the flowing material is measured and can be related to the volume fractions and dielectric constants of the component species as

$$\bar{D}_M = \epsilon D_1 + (1 - \epsilon)D_2 \tag{7-2}$$

with calibration, this unit can be used as a flow measuring device as well as an instantaneous voidage fraction meter.

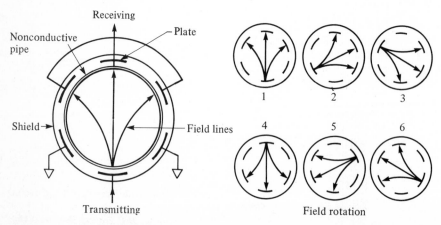

Figure 7-13 Auburn International capacitance meter. (Reprinted with permission of Auburn International.)

Nuclear Sources

Short-level radioactive sources are also being used as gas-solid flow measuring instruments. Generally, these units have slow response times and are not applicable for measurement in small-diameter pipes unless a special section can be created to look axially for the measurement.

Coriolis Force

A mass flowmeter based on the Coriolis force principle has been developed by Micro Motion. Example 7-3 describes some of its details.

Example 7-3 A Coriolis mass flowmeter uses a C-shaped pipe and a T-shaped leaf spring as opposite legs of a tuning fork (see Fig. 7-14). The T-shaped leaf spring is clamped to the stationary inlet/outlet end of the C-shaped flow pipe. A magnetic sensor/forcer coil is mounted on the leg of the leaf spring. A permanent magnet, suspended from the center of the C-shaped pipe, passes through the middle of the sensor/forcer coil.

The pipe and the leaf spring oscillate $180°$ out of phase with each other in the same way a tuning fork oscillates. The frequency of oscillation is determined by the natural frequency of the pipe/leaf spring. The period of a spring-mass system is given as

$$\tau = 2\pi \frac{M}{k_0}$$

where M = mass oscillating

k_0 = spring constant

The density of the gas-solid flow in such a system can be found from the period of the spring-mass arrangement:

$$M = M_{\text{system}} + M_{\text{fluid}}$$

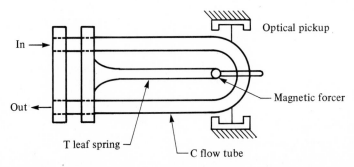

In →

Out ←

Optical pickup

Magnetic forcer

T leaf spring

C flow tube

Figure 7-14 Micro Motion mass flowmeter. (Reprinted with permission of Micro Motion, Inc.)

For the fluid:

$$M_{\text{fluid}} = \bar{V}_f \rho_f$$

For the meter:

$$M_{\text{system}} = \text{constant}$$

Thus,

$$\rho = 2\pi M_{\text{system}} + \bar{V}_f \rho_f$$

$$\frac{(\tau/2\pi) - M_{\text{system}}}{\bar{V}_f} = \rho_f$$

7-6 DESIGN CONSIDERATIONS

Whether transporting solids and gases together for industrial application or academic study, the design of an operative system is necessary. Some design suggestions and precautions will be considered here for a system that can be used with facility. Some suggestions have already been made on the use of a pressure tap installation in such a flow system. The use of these metal filters can have other applications in the gas-solid flow field. In sampling a flow system with positive or negative pressure, the exhaust port from the receivers can easily be equipped with these metal filters. They can be very effective, even down to the submicron ranges. This wide range of applicability makes them a very useful tool in handling dangerous and poisonous materials.

A system that works on a positive pressure flow supplied by a blower, fan, compressor, etc., is the generally preferred design. A negative pressure system can be utilized in the case of hazardous materials. In such systems, any leaks will suck in the surrounding gas rather than dispersing the solid matter into the work area.

Open- and closed-loop arrangements can be utilized. A closed-loop system has an advantage in that once the system is charged at certain conditions, additional feeding and continual recovery are not necessary. In studies on closed-loop systems the question of particle attrition as well as piping attrition must be raised. Generally, such closed-loop systems generate large electrostatic charges on the particles. Adequate grounding must be supplied to eliminate this charging effect unless one wishes to study the effect. In an open-loop system continuous collection must be provided, usually by a cyclone followed by a filter and let-down valve if the system is being operated at pressure. The attrition of particles and charging magnitudes are generally less in open-loop system designs.

A general comment is in order on electrostatics. As noted in chapter 6, if one wishes to eliminate these troublesome effects, the humidity of the transporting gas can be increased to 75% or more. Often during winter month operations, the air is at low relative humidity and the electrostatic charging is sizable, frequently increasing the overall energy losses by a sizable amount.

One of the most troublesome aspects of gas-solid flow design is devising a proper feeder. Blow tanks are simple devices for penumatic conveying of solids. Basically, the tank is filled with the solid and emptied through a pipeline by the expanding compressed gas admitted to the tank. These units can operate under a wide range of pressures and are known to convey materials up to 1 in. in diameter for several hundred feet. The blow tank operation generally is an intermittent one. In designing a blow tank system, care should be taken to have safety valves installed. Figure 7-15 shows a typical blow tank operation

Belt feeders and screw feeders normally operate at atmospheric pressure or under slight pressure. Some recent developments have produced pressurized screw feeders that can accurately deliver quantities of solid at prescribed rates. Feeding from a bin by gravity or higher pressures through an orifice mechanism or valve can also be used rather successfully at low pressures.

For orifice feeders, Tarman and Lee (1965) suggest the following model for the solid flow rate:

$$F' = \frac{0.0235\rho_B C_w}{D_p^{0.2}} \left(\frac{g}{\tan \beta}\right)^{0.5} D_o^{2.7} \tag{7-3}$$

where F' = flow rate, kg/hr
ρ_B = bulk density of solids, kg/m^3
C_w = wall correction factor
β = angle of repose
D_o = diameter of orifice, m
D_p = diameter of particles, m

The correction factor is given as

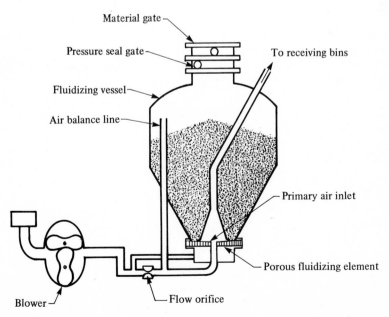

Figure 7-15 Blow tank arrangement (Kraus, 1968). Reproduced with permission of McGraw-Hill, Inc.

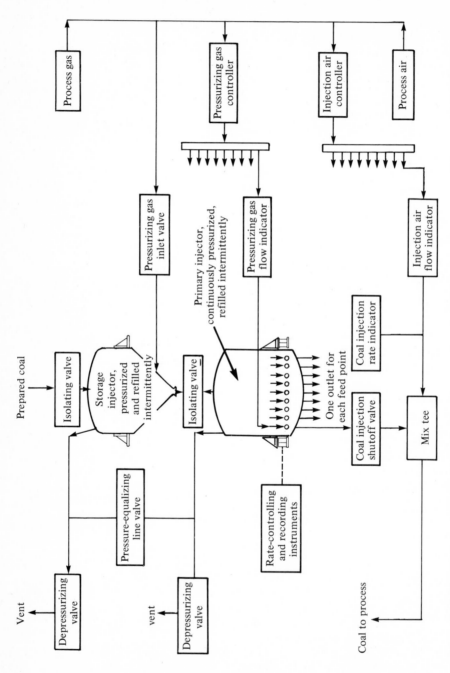

Figure 7-16 Simplified flow diagram of a typical Petrocarb continuous injection system for feeding against intermediate pressures. (Reproduced with permission of Petrocarb, H. Reintjes.)

$$C_w = \left(1 - \frac{D_p}{D_o}\right)^2 \tag{7-4}$$

For screw feeders, Tarman and Lee (1965) recommend

$$F' = 3600 \rho_B N_R \left\{ (\underline{D} - \underline{d}) \left[(\underline{D} + \underline{d}) \frac{\pi p'}{4} - \underline{t} \sqrt{\pi D_m^2 + P'^2} \right] \right\} \tag{7-5}$$

where N_R = rps

$\underline{D}$ = major screw diameter, m

$\underline{d}$ = minor screw diameter, m

$\underline{P'}$ = pitch, m

$\underline{t}$ = thickness of flight, m

$D_m = \dfrac{\underline{D} + \underline{d}}{2}$

Another feeder that has seen some use is the star-wheel feeder, which delivers solids from a rotating star-wheel.

Generally, lock hoppers have been employed for the troublesome high-pressure feeding of solids. In this type of system the solids are first fed to storage units (Fig. 7-16). As one hopper empties into the process, the second is held in reserve for the emptying of the first hopper. At this time the first one is shut and the second hopper feeds the system while the first hopper is refilled.

Another way to handle the high-pressure feeding operation is to create a slurry of the solids with a liquid and pump the slurry to the process. Ferretti (1976) compares the merits of the slurry and lock hopper feeds in Table 7-2.

Table 7-2 Advantages and disadvantages of solid feed systems (Ferretti, 1976)*

Slurry feed systems
Advantages
Simple to operate
Good reliability
Good where solvent is useful component of system
Shortcomings
Erosion at critical points (recirculating pump and injection valves)
Eenergy loss penalty where solvent must be evaporated from system

Pressurized lock hoppers
Advantages
Permits dry feed injection
No solvent dilutent
Shortcomings
Erosion at critical points (isolation valves)
Requires complex time cycle control combined with special hopper level control
Energy loss in pressurizing system
Cyclic operation

*Reprinted with permission of Coal Gasification Conference, University of Pittsburgh.

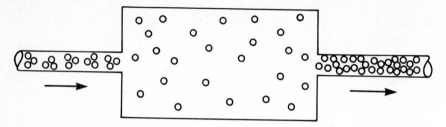

Figure 7-17 Expansion volume arrangement for gas-solid systems.

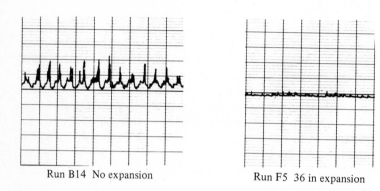

Run B14 No expansion

Run F5 36 in expansion

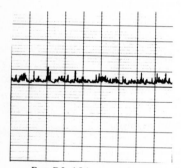

Run D2 15 in expansion

Figure 7-18 Flow fluctuations and associated damping; $W_s/W_g = 3.5$.

Some tests have been conducted on the use of an expansion volume as a capacitor in gas pipelines to produce steady solid flow. Figure 7-17 shows such an arrangement. The degree of damping of the solid fluctuating flow can be seen in Fig. 7-18 (Klinzing, 1978). Example 7-4 gives some details of the analysis on this flow dampener.

Example 7-4 Applying a material balance on the expansion volume seen in Fig. 7-17, one finds

$$\frac{d}{dt}[Nv_p^*\bar{V}\rho_p + \bar{V}(1 - Nv_p^*)\rho_f] = (U_p A Nv_p^*\rho_p)_{\text{in}} - U_p A Nv_p^*\rho_p$$

where v_p^* = volume of particle
$\bar{V}$ = volume of expansion unit

Assuming that the entrance number density fluctuates in a sinusoidal fashion,

$$N_{\text{in}} = N_o + N_i \cos \omega t$$

where N_i = amplitude of number desntiy fluctuation
N_o = mean value of the number density

This form can be substituted in the above to give

$$N = N_o + \frac{N_i}{\sqrt{1 + \omega^2 \alpha^2}} \cos(\omega t + \emptyset)$$

where

$$\alpha = \frac{\bar{V}(\rho_p - \rho_f)}{U_p \rho_p A}$$

From control theory, the term

$$\frac{1}{\{[1 + \omega^2 \bar{V}^2(\rho_p - \rho_f)^2]/U_p^2 \rho_p^2 A^2\}^{1/2}}$$

is the ratio of the amplitudes of output/input. This grouping then controls the degree of damping of the number density.

Comments have been made previously about the design of bends. Current thinking is to avoid broad, sweeping bends to eliminate particle-wall abrasion. Instead of these bends, a common T with one end plugged is recommended. Solids willl build up on the closed end of the T and the solids will abrade against themselves. Cleaning of these Ts is an easy operation, as mentioned earlier.

As a method of controlling the flow rate and giving an accounting of the rate, load cells have been incorporated into bin feeding systems. The bin feeder and receivers are often equipped with load cell arrangements of varying sensitivities. Digital readouts are available. Often, such systems are sensitive to external vibrations, so care must be taken to isolate the feeders and receivers from outside disturbances.

PROBLEMS

7-1 Perform a differential analysis on the instrumented 90° elbow arrangement in a 0.0254-m-diameter piping system. Consider the case where the internal pressure is 2026 N/m^2 and the solids loading is 3.5. Air is the transporting medium at a velocity of 30.5 m/sec. In order to note a 5% change in the force, what size changes must be present in the solids flow rate?

7-2 Assuming that solid flow fluctuations can be damped in an expansion section in the same manner as gas pulsation in compressor systems, determine the length of 0.0508-m-ID pipe needed to damp 25% of the incoming solid flow fluctuations. The solids are coal particles with a density of 1280 kg/m^3 and an average diameter of 150 μm, and they are flowing in a 0.0254-m-ID pipe in a gas stream with a density 1.2 kg/m^2 and velocity of 15.25 m/sec. The solid flow fluctuation is 10 cps and the particle velocity can be found by use of Hinkle's correlation.

7-3 In testing the pressure drop across a flow system carrying solids by a gas at 517×10^3 N/m^2, experiments have been run transferring coal from one receiver to another on a repeated basis. The data are given in the table. Plot the data and analyze the behavior seen. The average particle size of the initial coal particles is 200 μm.

ΔP, kN/m^2	Coal flow rate, kg/sec	ΔP, kN/m^2	Coal flow rate, kg/sec
24.1	0.0401 ± 0.0066	68.95	0.0946 ± 0.0068
31	0.0479 ± 0.0077	75.8	0.0993 ± 0.0074
41.4	0.0596 ± 0.0077	82.7	0.0995 ± 0.0114
48.3	0.0734 ± 0.0049	89.6	0.0993 ± 0.0069
55.2	0.0786 ± 0.0089	96.5	0.0964 ± 0.0285
66.1	0.0874 ± 0.0064	103	0.102 ± 0.0083

7-4 The Coriolis mass flowmeter works on the principle of the movement of a body on a revolving

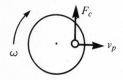

surface, as shown in the figure. The Coriolis force on a particle experiencing an angular velocity w is

$$F_c = 2m_p \omega v_p$$

where w = angular velocity
v_p = velocity of particle

Considering the C-shaped pipe arrangement (as is also seen in Fig. 7-14, it can be seen that for an oscillating C-shaped section, moments are set up:

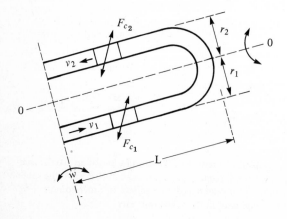

Write the moment about 0–0 using the Coriolis force expression and obtain an expression relating the total moment to the total mass flow rate through the tube.

The end view of the C-shaped pipe shows the deflection angle θ due to the moment.

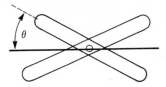

Assuming K_s is the angular spring constant of the pipe system, the moment can be related to this deflection angle. As seen in Fig. 7-14, the optical pickup can measure the time between pulses to give

$$\theta = \frac{L\,\omega\,\Delta t}{2r}$$

where Δt is the time between pulses. With this information, determine the mass flow rate in terms of Δt and the pipe geometry.

7-5 Mann and Crosby (1977) used a transducer to measure the collision rate of particles with its own surface in a gas-solid flow system. The number of collisions per net time at any radial position is proportional to the local flux of particles. Assume the constant of proportionality to be 10% of the cross-sectional area of the pipe. For a particular condition, it was found that the collision rate of particles I was

$$I \text{ collisions/sec} = \left[15 - 5\left(\frac{r}{r_o}\right)^{1/7} \right] \times 10^4$$

Using this information, determine the flow rate of particles for this condition in a 0.0508-m-diameter pipe.

REFERENCES

Babh, V. I., and V. B. Etkin: *Teploenergetika* 21(2):62–65, 84 (1974).

Baldwin, L. V., and W. R. Mickelson: *J. Eng. Mech. Div. Amer. Soc. Civil Eng.* 88:42 (1963).

Berman, N. S., and J. W. Dunning: *J. Fluid Mech.* 66:289 (1973).

Bobkowicz, A. J., and W. H. Gauvin: *Can. J. Chem. Eng.* 43:85 (1965).

Boothroyd, P. G.: *Flowing Gas Solids Suspensions,* Chapman and Hall, Ltd, London (1971).

——: *Trans. Inst. Chem. Eng.* 45:297 (1967).

Briller, R., and M. Robinson: *AIChE J.* 15:733 (1969).

Chao, B. T., H. Perez-Blanco, J. H. Saunders, and S. L. Soo: *Proc. Powder and Bulk Solids Conference,* Philadelphia (May 1979) A385. Published by Industry and Scientific Conference Management, Inc., Chicago.

Cheng, L., and S. L. Soo: *J. Appl. Phys.* 41:585 (1970).

Cheng, L., S. K. Tung, and S. L. Soo: *J. Eng. Power, Trans. ASME* 91:135 (Apr. 1970).

Dunning, J. W., and J. C. Angus: Chem. Eng. Sci. Div., Res. Rept. no. 10-04-67, Case Western Reserve University, Cleveland, 1967.

Ferretti, E. J.: *3d Annual Coal Gasification Conference,* University of Pittsburgh, 1976.

Gibson, H. G., W. T. Abel, and G. E. Fasching: *ASME Multiphase Symp.* 49 (1963).

Kane, R. S.: "Drag Reduction in Dilute Flowing Gas-Solid Suspension," Ph.D. thesis, City University, New York, 1973.

——, S. Weinbaum, and R. Pfeffer: *Pneumotransport* 2(BHRA):C3 (1973).

King, P. W.: *Pneumotransport* 2:D2 (Sept. 1973).

Klinzing, G. E.: *Chem. Eng. Sci.* **34**:971 (1979).

——: *IEC Process Des. Dev.* **19**:31 (1980).

Kolansky, M. S., S. Weinbaum, and R. Pfeffer: *Pneumotransport* **3** (1976).

Kraus, M. N.: *Pneumatic Conveying of Bulk Material,* Ronald Press, New York, 1968.

LeSage, L. G., and N. M. O'Fallon: *Quart. Tech. Rept., Argonne Natl. Lab* (Apr. 1977).

McCarter, R. J., L. F. Stutzman, and H. A. Koch: *Ind. Eng. Chem.* **41**:1290 (1949).

McLeman, M., and J. F. Richardson: *Trans. Inst. Chem. Eng.* **38**:257 (1960).

Mann, U., and E. J. Crosby: *IEC Process Des. Dev.* **16**:9 (1977).

Micro Motion (Boulder Colo.) Mass Flow Meter Sales Bulletin (1977).

Morrow, D. L., and J. C. Angus: Chem. Eng. Sci. Div. Res. Rept. no. 26-01-68, Case Western Reserve University, Cleveland, 1968.

Peskin, R. L.: *66th AIChE Meeting,* Philadelphia, 1973. p. 53.

Saltsman, R. D.: *Proc. Symp. IGT, BOM, Pneumatic Transportation of Solids,* Morgantown, W. Va. (Oct. 1965).

Schlinger, W. G., and B. H. Sage: *Ind. Eng. Chem.* **45**:657 (1953).

Soo, S. L., H. G. Ihrig, Jr., and A. F. El Kouch: *Trans. ASME J. Basic Eng.* **82D**:609 (1960).

Tarman, P. B., and B. S. Lee: *Proc. Symp. IGT, BOM, Pneumatic Transportation of Solids,* Morgantown, W. Va. (Oct. 1965), p. 32.

Towle, W. L., and T. K. Sherwood: *Ind. Eng. Chem.* **31**:457 (1939).

Van Zoonen, D.: *Proc. Symp. on the Interaction between Fluid and Particles,* Institute of Chemical Engineering, London, 1962.

Zlabin, V. V., and A. Z. Rozenshtein: *J. Appl. Mech. Tech. Phys.* **1**:192 (1975).

INDEX